Ramsey Theory

**WILEY SERIES IN
DISCRETE MATHEMATICS AND OPTIMIZATION**

A complete list of titles in this series appears at the end of this volume.

Ramsey Theory

Second Edition

Ronald L. Graham
Bruce L. Rothschild
Joel H. Spencer

Published by John Wiley & Sons, Inc., Hoboken, New Jersey.
Published simultaneously in Canada.

For general information on our other products and services please contact our Customer Care Department within the United States at (800) 762-2974, outside the United States at (317) 572-3993 or fax (317) 572-4002.

Wiley also publishes its books in a variety of electronic formats. Some content that appears in print, however, may not be available in electronic formats. For more information about Wiley products, visit our web site at www.wiley.com.

Library of Congress Cataloging-in-Publication Data is available.

ISBN 978-1-118-79966-6

10 9 8 7 6 5 4 3

Contents

Preface to First Edition

The classic theorems of Ramsey theory are known to many mathematicians, for there is an elegance in their statement. Van der Waerden: If the positive integers are finitely colored then one color class contains arithmetic progressions of arbitrary length. Schur: If the positive integers are finitely colored then one color class contains x, y, z with $x + y = z$. Ramsey: If a graph contains sufficiently many vertices (dependent on k) then it must contain either a complete set or an independent set of vertices of size k. The proofs are not so widely known. Our intent is to remedy this situation.

The origins of Ramsey theory are diffuse. Frank Ramsey was interested in decision procedures for logical systems. Issac Schur wanted to solve Fermat's last theorem over finite fields. B. L. van der Waerden solved an amusing problem—and immediately returned to his researches in algebraic geometry. The emergence of Ramsey theory as a cohesive subdiscipline of combinatorial analysis occurred only in the last decade. The central role of the Hales–Jewett theorem (the pure form of van der Waerden's theorem) has been recognized and exploited. The work of Walter Deuber (on the shoulders of Richard Rado), Jarik Nešetřil and Vojtech Rödl, Klaus Leeb, and others has given sharp definition to the subject. The field is alive and exciting. We indicate possible courses for future research but make no predictions.

In the first four chapters we attempt to give clear, self-contained expositions of the central results of Ramsey theory. The only requirement for the reader is that elusive "mathematical maturity." Chapter 5 deals on a more technical level with recent developments in the field. In the final chapter we explore the influence of outside disciplines, including the applications of topological dynamics spearheaded by Furstenberg and a combinatorial approach to the undecidability results of Paris and Harrington. There are general reference citations at the end of each of the first four chapters. In the last two chapters references are cited in the text.

We wish to make special acknowledgment of our debts to Paul Erdös, who provided us with constant encouragement and who can rightfully be considered the father of modern Ramsey theory, and to Ernst Straus, whose wisdom transcends the area of mathematics.

Finally, the junior author again wishes to thank his wife, Maryann, for her assistance, encouragement, and understanding. Without her, this enterprise would have had little meaning.

RONALD L. GRAHAM
BRUCE L. ROTHSCHILD
JOEL H. SPENCER

Murray Hill, New Jersey
Los Angeles, California
Stony Brook, New York
July 1980

Preface to the Second Edition

The romanticized view of mathematics is that it proceeds in sudden bursts of brilliant insight. Sometimes it happens just that way. Van der Waerden's theorem, the central result of Ramsey theory, was proven in 1926. As van der Waerden recalled:

> After lunch we went into Artin's office in the Mathematics Department of the University of Hamburg, and tried to find a proof. We drew some diagrams on the blackboard. We had what the Germans call "Einfälle": sudden ideas that flash into one's mind. Several times such new ideas gave the discussion a new turn, and one of the ideas finally led to the solution.
> [van der Waerden 1971]

Van der Waerden's proof used a subtle double induction and when expressed quantitatively led to an extremely fast growing function. Mathematicians—we three included—searched for a different proof technique without these features. In 1987 Saharon Shelah was shown van der Waerden's theorem and within a day or two found a new proof. Whether *Einfälle* or not, Shelah's proof avoids the double induction, involves only "reasonably" fast growing functions, and—best of all—is totally elementary. In this edition we give a complete treatment of Shelah's proof as well as the original proof of van der Waerden.

The response to the first edition of this volume has been most gratifying. Before its publication this subject matter had been generally regarded as a collection of loosely tied results. Today it is recognized for what it is—a cohesive subdiscipline of Discrete Mathematics. We are particularly pleased with the name given to this subdiscipline: Ramsey theory!

RONALD L. GRAHAM
BRUCE L. ROTHSCHILD
JOEL H. SPENCER

Murray Hill, New Jersey
Los Angeles, California
New York, New York
July, 1989

Preface to the Paperback Edition

We have long felt that Ramsey Theory was one of the gems of mathematics. Inside any large structure, be it random, capricious or adversarial, lies a highly ordered small structure. Ramsey's Theorem, van der Waerden's Theorem, the Hales-Jewett Theorem, and Szemerédi's Theorem are now recognized as vital insightful results. We like to feel that this volume has played a modest role in the recognition that Ramsey Theory is more than a collection of beautiful theorems but"rather a "theory," with distinct methodologies.

The entire area of Discrete Mathematics has had a remarkable rise in the Mathematical Universe over the past few decades. No longer is it regarded as "the slums of topology," or worse. The awarding of the Kyoto Prize to László Lovász (2010), the awarding of the Fields Medals to Timothy Gowers (1998) and Terrace Tao (2006) are indicators of the respect paid to Discrete Mathematics. We take particular note on the awarding of the Abel Prize (2012) to Endre Szemérdi, whose work may be regarded as combinatorial in its entirety

We now mark the centennial of the birth of Paul Erdös (1913-1996). Uncle Paul, as his myriad friends and collaborators called him, was a giant of twentieth century mathematics. Ramsey Theory is correctly named after Frank Plumpton Ramsey, who first proved the eponymous Ramsey's Theorem. However, In terms of the development of the field it would be more accurate to call it Erdös Theory. Erdös became interested in the area while still a teenager. In the winter of 1931–2 (the paper was published in 1935) he worked with George Szekeres and Esther Klein on a geometric problem described in §1.7. He had numerous papers on all aspects of Ramsey Theory, from the asymptotic bounds on the Ramsey Function to Restricted Ramsey Results to Canonical Ramsey Theorems and much much more. More than six decades after his original infatuation he wrote a paper (1995) with Stephen Suen and Peter Winkler that played a key role in the asymptotic determination of the Ramsey Function $R(3, k)$\$. Erdös's influence should not be measured solely by his theorems, great as they are. Erdös was an inspirational figure. Fan Chung describes the feeling of collaboration with Erdös well:

Working with Paul Erdös was like taking a walk in the hills. Every time when I thought that we had achieved our goal and deserved a rest, Paul pointed to the top of another hill and off we would go.

We end with Paul Erdös's constant admonition:

Prove and Conjecture!

THE AUTHORS

Notation

A few specialized notations are particularly useful throughout Ramsey theory. We give them here.

$N = \{1, 2, \ldots\}$ = the positive integers.
$|X|$ = cardinality of X.
$[n] = \{1, \ldots, n\}$, defined for $n \in N$. Often we use $[n]$ when we wish to refer to an arbitrary set of cardinality n.
$[X]^k = \{Y \colon Y \subset X, |Y| = k\}$.
$[X]^{\leq k} = \{Y \colon Y \subset X, |Y| \leq k\}$.
$[X]^{<\omega} = \{Y \colon Y \subset X, Y \text{ finite}\}$.

When $X = [n]$ we remove the second set of brackets. Thus:

$$[n]^k = \{Y \colon Y \subset \{1, \ldots, n\}, |Y| = k\}.$$

We write $\{x_1, \ldots, x_n\}_<$ for a set $\{x_1, \ldots, x_n\}$ such that $x_1 < \cdots < x_n$. If χ is a map with domain $[A]^k$ we write $\chi(a_1, \ldots, a_k)$ for $\chi(\{a_1, \ldots, a_k\})$ when there is no danger of confusion.

K_n denotes a complete graph on n points.

Arithmetic Progression is abbreviated AP.

The Pigeon-Hole principle: If m pigeons roost in n holes and $m > n$ then at least two pigeons must share a hole. More prosaically: If m objects are colored with n colors and $m > n$ then some two objects have the same color.

1

Sets

"Of three ordinary people, two must have the same sex."

D. J. Kleitman

1.1 RAMSEY'S THEOREM ABRIDGED

In any collection of six people either three of them mutually know each other or three of them mutually do not know each other.

This "puzzle problem" may be considered the first nontrivial example of what we shall call Ramsey theory. We begin this volume with an expository proof of this result.

We have tacitly assumed that the relation of "knowing" is symmetric; that is, if A knows B then B knows A. We do *not* assume transitivity; if A knows B and B knows C then A may or may not know C.

Fix one person, say A, and consider his or her relation to the other five, say B, C, D, E, and F. He or she must either know at least three of them or not know at least three of them ($2 + 2 < 5$). Suppose that A knows three of them, say C, E, and F. If some pair of these three, say C and F, know each other then A, C, and F are three people who mutually know each other. If no pair of the three know each other then those three mutually do not know each other. In either case we have found a threesome with the desired property. Of course, if A does not know three of the others the argument is identical.

As this is a mathematics book it will be necessary to adopt some formalisms. An *r-coloring* of a set S is a map

$$\chi: S \to [r].$$

For $s \in S$, $\chi(s)$ is called the color of s. We say that a set $T \subseteq S$ is monochromatic (under χ) if χ is constant on T.

At this point we introduce the *arrow notation*, which has proved particularly useful in Ramsey theory.

DEFINITION.   We write

$$n \to (l)$$

if, given any 2-coloring of $[n]^2$, there is a set $T \subseteq [n]$, $|T| = l$ so that $[T]^2$ is monochromatic. When $[T]^2$ is monochromatic, $|T| = l$, T is called a monochromatic K_l.

Our original problem is equivalent to the assertion $6 \to (3)$. Identify the six people $A, \ldots, F$ with elements $1, \ldots, 6$, respectively, and identify the statement "A and B know (do not know) each other" with the relation $\chi(1, 2) = 1$ ($\chi(1, 2) = 2$). We shall, of course, work with the more mathematical format. The arrow relation is generalized in a number of ways.

DEFINITION.   We write $n \to (l_1, \ldots, l_r)$ if, for every r-coloring of $[n]^2$, there exists i, $1 \leqslant i \leqslant r$, and a set $T \subseteq [n]$, $|T| = l_i$ so that $[T]^2$ is colored i.

For example, $10 \to (4, 3)$ is the assertion that, given any ten people, either four of them mutually know each other or three mutually do not know each other. In the case $l_1 = \cdots = l_r = l$ we use the shorthand

$$n \to (l)_r,$$

that is, every r-coloring of $[n]^2$ yields a monochromatic $[l]^2$. If the number of colors r is not indicated it is assumed to be 2. Thus $n \to (l)$, $n \to (l)_2$, and $n \to (l, l)$ denote the same thing.

We note the following important trivialities.

1. If $l_i' \leqslant l_i$, $1 \leqslant i \leqslant r$, and $n \to (l_1, \ldots, l_r)$ then $n \to (l_1', \ldots, l_r')$.
2. If $m \geqslant n$ and $n \to (l_1, \ldots, l_r)$ then $m \to (l_1, \ldots, l_r)$.
3. Let σ be a permutation of $[r]$. Then $n \to (l_1, \ldots, l_r)$ iff $n \to (l_{\sigma 1}, \ldots, l_{\sigma r})$.
4. $n \to (l_1, \ldots, l_r)$ iff $n \to (l_1, \ldots, l_r, 2)$. In particular, $l_1 \to (l_1, 2)$.

To illustrate, the statement $10 \to (4, 3)$ logically implies $10 \to (3, 3)$, for if four people mutually know each other, one may be deleted. Similarly, $10 \to (4, 3)$ implies $11 \to (4, 3)$ since the eleventh person may be ignored. If $10 \to (4, 3)$ then $10 \to (3, 4)$, as we may interchange the roles of knowing and not knowing. Finally, if $10 \to (4, 3)$ then $10 \to (4, 3, 2)$, for suppose that every pair is either loving, hating or avoiding—and these categories are mutually exclusive! One avoiding pair will form the desired monochromatic 2-set. If no pair is avoiding then all pairs are loving or hating so that $10 \to (4, 3)$ implies the existence of the desired monochromatic set.

DEFINITION. The *Ramsey function* $R(l_1, \ldots, l_r)$ denotes the minimal n such that

$$n \to (l_1, \ldots, l_r).$$

We let $R(l; r)$ denote $R(l_1, \ldots, l_r)$, where $l_1 = \cdots = l_r$ and $R(l) = R(l; 2) = R(l, l)$.

From trivialities 1–4 we note that $R(l_1, \ldots, l_r)$ is monotonic in each variable and totally symmetric, $R(l_1, \ldots, l_r, 2) = R(l_1, \ldots, l_r)$, and $R(l, 2) = l$.

Theorem 1 (Ramsey's Theorem—Abridged). The function R is well defined; that is, for all $l_1, \ldots, l_r$ there exists n so that

$$n \to (l_1, \ldots, l_r).$$

We first give two proofs of this theorem for the case $r = 2$.

Proof 1. Use a double induction on l_1 and l_2. Note that $R(l, 2) = R(2, l) = l$ by triviality 4. Now assume, by induction, that $R(l_1, l_2 - 1)$ and $R(l_1 - 1, l_2)$ exist.

Claim. $R(l_1, l_2 - 1) + R(l_1 - 1, l_2) \to (l_1, l_2)$.

Proof. Fix a 2-coloring χ of $[n]^2$, $n = R(l_1, l_2 - 1) + R(l_1 - 1, l_2)$. Fix one element $x \in [n]$ and set

$$I_x = \{y \in [n]: \chi(x, y) = 1\},$$
$$II_x = \{y \in [n]: \chi(x, y) = 2\},$$
$$= [n] - I_x - \{x\}.$$

Then $|I_x| + |II_x| = n - 1$ so that either
  (a) $|I_x| \geqslant R(l_1 - 1, l_2)$
or
  (b) $|II_x| \geqslant R(l_1, l_2 - 1)$.
Assume (a). By the definition of R either there exists $T \subseteq I_x$, $|T| = l_2$,

such that $[T]^2$ is colored 2 (which is as desired) or there exists $S \subseteq I_x$, $|S| = l_1 - 1$, so that $[S]^2$ is colored 1. In the latter case set $S^* = S \cup \{x\}$. (Here is the critical point of the proof. Since $S \subseteq I_x$ all $\{x, s\}$, $s \in S$, are colored 1.) Then $|S^*| = l_1$ and $[S^*]^2$ is colored 1, as desired. Case (b) is symmetric.

[It may help the reader to see the following expository proof that $10 \to (4, 3)$. Consider any group of ten people (Fig. 1.1) Any one of these ten, say J, either knows at least six or does not know at least four of the remaining nine people. If J knows six then of those six either three know each other or three do not. In the former case, these three together with J are four who know each other. If there are four people J does not know either two of them do not know each other, and together with J make three, or the four mutually know each other.]

Proof 2. We show directly that

$$2^{2l-1} - 1 \to (l) \,.$$

Fix S_1, $|S_1| \geqslant 2^{2l-1} - 1$ and a 2-coloring χ of $[S_1]^2$. Define, for $1 \leqslant i \leqslant 2l - 1$, sets S_i and elements $x_i \in S_i$ as follows (Fig. 1.2):

(i) Having chosen S_i, select $x_i \in S_i$ arbitrarily.

Figure 1.1 $10 \to (4, 3)$.

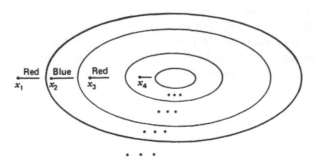

Figure 1.2 Proof 2 of Ramsey's theorem.

(ii) Having selected $x_i \in S_i$, set

$$T_j = \{u \in S_i : \chi(x_i, u) = j\}, \qquad j = 1, 2.$$

Set S_{i+1} equal to the larger (in cardinality) of T_1, T_2. Since $|T_1| + |T_2| = |S_i| - 1$, $|S_{i+1}| \geqslant (|S_i| - 1)/2$.

Since $|S_1|$ was sufficiently large we may select $x_1, \ldots, x_{2l-1}$ before this procedure terminates (when $S_i = \varnothing$). We define a new coloring

$$\chi^* : \{x_1, \ldots, x_{2l-1}\} \to \{1, 2\}.$$

Let $\chi^*(x_i)$ be that j (equal to 1 or 2) such that $\chi(x_i, y) = j$ for $y \in S_{i+1}$. Since the coloring χ^* splits the $2l - 1$ points into two groups we can find l points $x_{i_1}, \ldots, x_{i_l}$ so that

$$\chi^*(x_{i_s}) = j \qquad \text{for } 1 \leqslant s \leqslant l.$$

Then, for any $1 \leqslant s < t \leqslant l$, $x_{i_t} \in S_{i_t} \subseteq S_{i_s+1}$ and $\{x_{i_1}, \ldots, x_{i_l}\}$ is the desired monochromatic l-set.

A proof for an arbitrary number of colors r can be given along the lines of either of the preceding proofs. One may replace Proof 1 by an induction on $l_1, \ldots, l_r$ showing

$$2 + \sum_{i=1}^{r} R(l_1, \ldots, l_i - 1, \ldots, l_r) - 1 \to (l_1, \ldots, l_r)$$

or replace Proof 2 by the result

$$r^{(l-1)r+1} - 1 \to (l : r).$$

The details are left to the reader.

Our Proof 2 embodies a basic method of proof that we shall encounter frequently. We shall call it the Induced Coloring method. We 2-color the subobjects of a certain type of large structure S. Here S is the complete graph on n points, and the subobjects are edges—more generally, we call them snargles. We find a substructure T, perhaps much smaller than S but still very large, on which the coloring is canonical in some sense. There will be a subobject—let us call it a turble—so that the coloring of the snargles of T depends only on the leading turple contained in the snargle. (The turbles will be ordered explicitly.) In our example, turble = point and on T the coloration of $\{x, y\}_<$ depends only on x. Now we define a coloring of the turbles of T, coloring a turble by the color of all the snargles of which it is the leading turble. Assume that at some previous point we have proved a Ramsey theorem for turbles. Then we know there is a substructure U of T that is still large on which all the turbles are the same color in the induced coloring. But then, in the original colorings, all of the snargles of U have the leading turble in U and hence all are the same color.

The reader might note that Category theory could provide an appropriate vocabulary for the preceding discussion. Indeed, a number of authors have approached Ramsey theory from a Category theory point of view with some impressive results. In this volume, however, we have consciously attempted to avoid Category theory notation. We do this both for reasons of personal preference and in an attempt not to limit the readership of this book.

A detailed analysis of Proofs 1 and 2 shows that Proof 1 gives a better upper bound on $R(n)$ than Proof 2. In Chapter 4 we take a detailed look at the value of the Ramsey function and associated functions. However, outside of that chapter, we are generally interested not in the exact values of the Ramsey functions but rather in their existence. hence we usually shall not employ methods such as Proof 1. We shall sacrifice an improvement in the upper bounds on these functions in favor of clarity of exposition of the proofs of these existence theorems. As we shall see in Section 2.7, the functions associated with the theorems of later sections (e.g., van der Waerden's theorem or the Hales–Jewett theorem) apparently increase so rapidly that no moderate upper bound is currently known for them. This has been remedied substantially by the recent results of Gowers [2001].

The existence of $n = R(a, b)$ has a special interpretation in Graph theory terms. Let G be an arbitrary graph on n vertices. We may associate G with the 2-coloring of the complete graph K_n obtained by coloring $\{i, j\}$ red iff $\{i, j\}$ is an edge of G, and blue otherwise. Clearly this gives a bijective correspondence between 2-colorings of K_n and graphs G on n vertices. We rephrase Ramsey's theorem in the case as follows:

Any graph G on $n = R(a, b)$ vertices contains either a clique on a vertices or an independent set of b vertices.

1.2 RAMSEY'S THEOREM UNABRIDGED

We now consider colorations of $[n]^k$, where k is an arbitrary integer. This generalizes the case $k = 2$ of Section 1.1.

DEFINITION. $n \to (l_1, \ldots, l_r)^k$ if, for every r-coloring of $[n]^k$, there exists i, $1 \leq i \leq r$, and a set T, $|T| = l_i$ so that $[T]^k$ is colored i.

In the case $l_1 = \cdots = l_r = l$ we use the shorthand

$$n \to (l)_r^k .$$

We say in this case that every r-coloring of $[n]^k$ yields a monochromatic $[l]^k$. If the number of colors r is not indicated it is assumed to be 2. Thus $n \to (l)^k$, $n \to (l)_2^k$, and $n \to (l, l)^k$ are identical relations. This is consistent with previous notations—if k is not given it is also assumed to be 2.

The Ramsey function for k-sets is indicated by R_k:

$$R_k(l_1, \ldots, l_r) = \min\{n_0: \text{ for } n \geq n_0, n \to (l_1, \ldots, l_r)^k\} ,$$

$$R_k(l; r) = \min\{n_0: \text{ for } n \geq n_0, n \to (l)_r^k\} ,$$

$$R_k(l) = \min\{n_0: \text{ for } n \geq n_0, n \to (l)^k\} .$$

Theorem 2 (Ramsey's Theorem). The function R is well defined; that is, for all $k, l_1, \ldots, l_r$ there exists n_0 so that, for $n \geq n_0$,

$$n \to (l_1, \ldots, l_r)^k .$$

Proof. We use induction on k, following the lines of Proof 2 for each k. For $k = 1$ Ramsey's theorem becomes a triviality. We have

$$1 + \sum_{i=1}^{r} (l_i - 1) \to (l_1, \ldots, l_r)^1 ,$$

that is, if $n \geq 1 + \sum_{i=1}^{r} (l_i - 1)$ elements are r-colored, some color i is used at least l_i times. This is a general form of the Pigeon-Hole principle, defined earlier under "Notation," and Ramsey's theorem is often considered a generalization of it. For $k = 2$ we have already proved Ramsey's theorem, though the induction argument includes that case.

Assume that the result holds for $k - 1$; it suffices to find n so that

$$n \to (l)_r^k .$$

Basically, the k-element subsets become snargles, and the $(k-1)$-element subsets become turbles, as in our general discussion following Proof 2 of the Abridged Ramsey theorem.

Let n be "sufficiently large" (more on that later), and fix an r-coloring χ of $[n]^k$. Set $t = R_{k-1}(l; r)$, which exists by induction. Select distinct elements $a_1, \ldots, a_{k-2} \in [n]$ arbitrarily, and define $S_{k-2} = [n] - \{a_1, \ldots, a_{k-2}\}$.

Now we select a_i, S_i as follows:

(i) S_i having been defined, we select $a_{i+1} \in S_i$ arbitrarily.
(ii) Having selected a_{i+1}, we split $S_i - \{a_{i+1}\}$ into equivalence classes by

$$x \equiv y \text{ iff for every } T \subseteq \{a_1, \ldots, a_{i+1}\}, \quad |T| = k - 1,$$

$$\chi(T \cup \{x\}) = \chi(T \cup \{y\}) .$$

The equivalence class is therefore determined by the color of $\binom{i+1}{k-1}$ sets so there are at most $r^{\binom{i+1}{k-1}}$ such classes. We define S_{i+1} as the largest of those classes. Hence $S_{i+1} \subseteq S_i - \{a_{i+1}\}$ and

$$|S_{i+1}| \geq (|S_i| - 1) r^{-\binom{i+1}{k-1}} .$$

We choose n sufficiently large so that the procedure may be continued until a_t is defined. For definiteness, we select n so that the sequence with initial condition $u_{k-2} = n - (k - 2)$ and recursion

$$u_{i+1} = (u_i - 1) r^{-\binom{i+1}{k-1}}$$

satisfies $u_t \geq 1$. Certainly

$$n = 2r^c , \quad c = \sum_{i=k-1}^{t-1} \binom{i+1}{k-1}$$

will suffice. (The calculation is important for bounding the Ramsey function—see Section 4.7—but not for proving Ramsey's theorem.)

We now restrict our attention to the sequence $a_1, \ldots, a_t$. Suppose that $1 \le i_1 < i_2 < \cdots < i_{k-1} < s \le t$. Then $a_s \in S_{s-1} \subseteq S_{i_{k-1}+1}$. The color $\chi(\{a_{i_1}, \ldots, a_{i_{k-1}}, a_s\})$ remains the same if a_s is replaced by any $x \in S_{i_{k-1}+1}$ (by the definition of the equivalence classes), including any $x = a_r$, $k - 1 < r < t$. We define a coloring χ^* on $(k-1)$-subsets of $\{a_1, \ldots, a_t\}$ by

$$\chi^*(\{a_{i_1}, \ldots, a_{i_{k-1}}\}) = \chi(\{a_{i_1}, \ldots, a_{i_{k-1}}, a_s\})$$

for all $i_{k-1} < s \le t$. (A technical point: when $i_{k-1} = t$ we define χ^* arbitrarily.) By the definition of t there is a subsequence $\{b_1, \ldots, b_l\}$ of $\{a_1, \ldots, a_t\}$, which is monochromatic under χ^*—say that all $(k-1)$-subsets are red. Then, for any $1 \le j_1 < \cdots < j_{k-1} < j_k \le l$,

$$\chi(\{b_{j_1}, \ldots, b_{j_{k-1}}, b_{j_k}\}) = \chi^*(\{b_{j_1}, \ldots, b_{j_{k-1}}\})) = \text{red},$$

and so $\{b_1, \ldots, b_l\}$ is the desired monochromatic l-set.

1.3   VIEWS OF RAMSEY THEORY

We are concerned here with "Ramsey-type theorems." Rather than formally define this concept we state six major theorems and then consider their similarities. Formal definitions, proofs, and detailed discussion for Results 2–6 are given in later chapters.

Super Six

1. **Ramsey's theorem** (Sections 1.1 and 1.2): For all l, r, k there exists n_0 so that, for $n \ge n_0$, if $[n]^k$ is r-colored there exists a monochromatic $[l]^k$.

2. **Van der Waerden's theorem** (Section 2.1): For all l, r there exists n_0 so that, for $n \ge n_0$, if $[n]$ is r-colored there exists a monochromatic arithmetic progression $\{a, a + d, \ldots, a + (l-1)d\} \subseteq [n]$ of length l.

3. **Schur's theorem** (Section 3.1): For all r there exists n_0 so that, for $n \le n_0$, if $[n]$ is r-colored there exist $x, y, z \in [n]$, having the same color, so that

$$x + y = z.$$

A system of equations $\mathfrak{L}$ on variables $x_1, \ldots, x_m$ is called *regular* if, for all r, there exists n_0 so that, for $n \geq n_0$, if $[n]$ is r-colored there exist $x_1, \ldots, x_m \in [n]$, all the same color, satisfying $\mathfrak{L}$.

4. Rado's theorem (Section 3.2): The single equation

$$c_1 x_1 + \cdots + c_m x_m = 0$$

is regular iff some nonempty subset of the c_i sums to zero.

5. Hales–Jewett theorem (Section 2.2): For all r, k there exist n_0 so that, for $n \geq n_0$, if the n-dimensional cube

$$\{(x_1, \ldots, x_n): x_i \in \{0, 1, \ldots, k-1\}, 1 \leq i \leq n\}$$

is r-colored there exists a monochromatic "line."

6. Graham–Leeb–Rothschild theorem (Section 2.4): Fix a finite field F on q elements. For all k, l, r there exists n_0 so that the following holds for $n \geq n_0$. Let V be an n-dimensional vector space over F. Color the k-dimensional subspaces of V with r colors. Then there exists an l-dimensional subspace of V all of whose k-dimensional subspaces have the same color.

Segments of Ramsey theory may be described in the language of Lattice theory. Let $L_1 \subset L_2 \subset \cdots \subset L_n \subset \cdots$ be a sequence of graded lattices with a rank function denoted by ρ. The sequence is called Ramsey if, for all c, k, l, there exists n_0 so that, for $n \geq n_0$, if $\{x \in L_n: \rho(x) = k\}$ is c-colored there exists $y \in L_n$, $\rho(y) = l$ so that

$$\{x \in L_n: \rho(x) = k, x \leq y\}$$

is monochromatic. For the original theorem of Ramsey, L_n is the Boolean lattice of subsets of an n-element set with $\rho(A) = |A|$. For the Graham–Leep–Rothschild theorem, L_n is the subspace lattice of an n-dimensional vector space over a fixed finite field F and $\rho(V)$ is the dimension of V.

One might also view portions of Ramsey theory as statements about certain bipartite graphs (Fig. 1.3). A bipartite graph G consists of two sets T (top) and B (bottom) and a family $E(G)$ of edges $\{t, b\}$, $t \in T$, $b \in B$. We call G r-Ramsey if, given any r-coloring of B, there exists $t \in T$ such that

$$\{b \in B: \{t, b\} \in E(G)\}$$

is monochromatic. We call a sequence $\{G_i\}$ Ramsey if, for all r, there exist n_0 so that, for $n \geq n_0$, G_n is r-Ramsey.

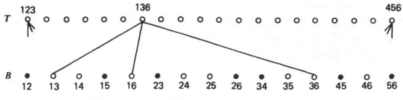

Figure 1.3 $6 \rightarrow (3)$ in bipartite graph format.

For fixed $k \leq l$ and lattice L_i we may generate a bipartite graph G_i by restricting our attention to ranks k, l, Formally, set

$$T = \{y \in L_i, r(y) = l\} ,$$
$$B = \{x \in L_i, r(x) = k\} ,$$
$$E(G_i) = \{\{x, y\}, x \in B, y \in T, x \leq y\} .$$

Then $\{L_i\}$ is Ramsey exactly when, for all k, l, the corresponding $\{G_i\}$ is Ramsey.

Van der Waerden's theorem may be expressed in this terminology. For a given l define g_n by

$$B = [n] ,$$

$T = $ the family of arithmetic progressions

$$S = \{a, a + d, \ldots, a + (l - 1)d\} \subseteq B ,$$
$$E(G)_n = \{\{x, S\}: x \in B, S \in T, x \in S\} .$$

Then $\{G_n\}$ is Ramsey.

A third approach to Ramsey theory utilizes the language of hypergraphs. A hypergraph H consists of a vertex set $V(H)$ and a family $E(H)$ of subsets of $V(H)$. The elements $X \in E(H)$ are called hyperedges. An *r-coloration* of H is a map

$$\chi: V(H) \to [r]$$

such that *no* $X \in E(H)$ is monochromatic. The *chromatic number* $\chi(H)$ of the hypergraph is the minimal r such that an r-coloration of H exists. We note that if all $X \in E(H)$ have $|X| = 2$ the hypergraph reduces to our usual concept of a graph, and chromatic number is as usually defined.

All "Ramsey theory" results may be expressed in hypergraph terminology. Let us take the Hales–Jewett theorem as an example. For a given k, n we may construct a hypergraph $H = H_{n,k}$ with vertex set $V(H) = C_k^n$, the n-dimensional cube over $[k]$, and $E(H)$ equal to the set of "lines" in C_k^n. Then, for all k, r, if n is sufficiently large $\chi(H_{n,k}) \geqslant r$. In more concise form:

$$\lim_n \chi(H_{n,k}) = +\infty \text{ for every } k.$$

1.4 RAMSEY THEOREMS AND DENSITY THEOREMS

DEFINITION. Let $H = (V, E)$ be a hypergraph. We define the *Turán function* $T(H)$ as the minimal T such that any set of vertices of cardinality at least T necessarily contains a hyperedge. We set $\tau(H) = T(H)/V(H)$.

Paul Turán found the exact value for $T(H)$, where $H = ([n]^2, \{[S]^2 : S \in [n]^k\})$. Here, in classical Graph theory terminology, $T(H)$ is the minimal number of edges on n points that ensure a clique on k points.

For a hypergraph $H = (V, E)$ we consider the following statements:

$$A: \chi(H) > r,$$

$$B: \tau(H) \leqslant r^{-1}.$$

For a sequence of hypergraphs $H_n = (V_n, E_n)$ we have the analogous statements:

$$A^*: \chi(H_n) \to +\infty \qquad \text{as } n \to \infty,$$

$$B^*: \tau(H_n) \to 0 \qquad \text{as } n \to \infty.$$

Statements B and B^* are *density* statements; B says that any sufficiently large set of vertices contains a hyperedge. Statements A and A^* are Ramsey statements; A says that if the vertex set is *partitioned* into r classes one class contains a hyperedge.

Theorem 3. (i) B implies A. (ii) B^* implies A^*.

Proof
- (i) Assume B. Let χ be an r-coloring of V. Some color must have been used on at least r^{-1} of the vertices. *That* color contains a hyperedge.
- (ii) Assume B^*. For a given r there exists n_0 so that $\tau(H_n) \leqslant r^{-1}$ for $n \geqslant n_0$. Thus $\chi(H_n) \geqslant r$ for $n \geqslant n_0$—hence A^*.

The converses, A implies B and A^* implies B^*, are *false*. Consider their interpretation for some of our basic Ramsey theorems.

For fixed l consider the sequence $H_n = (V_n, E_n)$, where $V_n = [n]^2$, $E_n = \{[S]^2 : S \in [n]^l\}$. Here $\chi(H_n) > r$ is identical to $n \to (l)_r$. By Ramsey's theorem (for $k = 2$) the sequence $\{H_n\}$ satisfies A^*. The classical Turán's theorem states that the maximal graph on n vertices without an l-cliquie is achieved by splitting the n vertices into $l - 1$ sets of cardinalities $[n/(l-1)]$ and $[n/(l-1)] + 1$ and placing an edge between any pair of vertices in different sets. In our terminology, if

$$n = (l-1)m + r, \qquad 0 \leqslant r < l - 1,$$

$$T(H_n) = 1 + \binom{n}{2} - r\binom{m+1}{2} - (l - 1 - r)\binom{m}{2}$$

and

$$\lim_{n \to \infty} \tau(H_n) = 1 - \frac{1}{l-1}$$

so that B^* is false!

Van der Waerden's theorem may also be interpreted in this light. For fixed l the statements A^*, B^* become as follows:

A^*: For all r there exists $n_0(r)$ so that if $n \geqslant n_0(r)$ and $[n]$ is r-colored there exists a monochromatic arithmetic progression of length l.

B^*: For all $\varepsilon > 0$ there exists $n_0(\varepsilon)$ so that if $n \geqslant n_0(\varepsilon)$ and $S \subseteq [n]$, $|S| \geqslant n\varepsilon$ then S contains an arithmetic progression of length l.

Van der Waerden's theorem, A^*, was proved in 1927. Statement B^* was conjectured by P. Erdös and P. Turán in 1936. The case $l = 2$ is trivial, $l = 3$ was settled positively by K. Roth in 1952, and $l = 4$ by E. Szemerédi in 1969. Finally, in 1973, Szemerédi proved B^* for all l; this result is discussed in Section 2.5. The full proof is highly complex, is supremely ingenious, and is by no means a "simple corollary" of A^*.

1.5 THE COMPACTNESS PRINCIPLE

In most of our Ramsey theorems we prove that, for n sufficiently large, an r-coloring of $[n]$ (or $[n]^k$) has a certain property. In this section we show that it is often sufficient to prove that any r-coloring of N or $[N]^k$ has the property.

DEFINITION. Let $H = (V, E)$ be a hypergraph, $W \subseteq V$. The restriction of H to W, denoted by H_W, is the hypergraph $H_W = (W, E_W)$, where

$$E_W = \{X \in E : X \subseteq W\}.$$

Theorem 4 (Compactness Principle). Let $H = (V, E)$ be a hypergraph where all $X \in E$ are finite (but V need not be). Suppose that, for all $W \subseteq V$, W *finite*,

$$\chi(H_W) \leq r.$$

Then

$$\chi(H) \leq r.$$

The theorem is often expressed in contrapositive form: *If $\chi(H) > r$ there exists a finite W such that $\chi(H_W) > r$.*

We give two proofs. The first proof is for the case V countable. (The case V finite is tautological—take $W = V$.) The second proof works for arbitrary V but requires the Axiom of Choice (in fact, the Compactness principle cannot be proved from the usual axioms of set theory without the Axiom of Choice).

Proof 1. We assume that V is countable in this proof. Our proof is essentially a diagonal argument. For convenience consider $V = N$. For all $n \in N$ there exists a coloring

$$\chi_n : [n] \rightarrow \{1, \dots, r\}$$

so that no $A \in E$, $A \subseteq [n]$, is monochromatic. We define a function

$$\chi^* : N \rightarrow \{1, \dots, r\}$$

by induction. We assume that $\chi^*(1), \dots, \chi^*(j-1)$ have been defined so that

$$S_{j-1} = \{n : n \geq j - 1 \text{ and } \chi^*(i) = \chi_n(i) \text{ for } 1 \leq i \leq j - 1\}$$

is infinite. We partition $S_{j-1} - \{j-1\}$ ($j-1$ may or may not be in S_{j-1}) into r classes, depending on the value of $\chi_n(j)$. For some color c,

$$T = \{n \in S_{j-1} : \chi_n(j) = c\}$$

is infinite. Then we set $\chi^*(j) = c$ and $S_j = T$.

We claim that χ^* is the desired r-coloring of H. Let $X = \{x_1, \ldots, x_m\}_< \in E$. Since $S_{x_m} \neq \varnothing$ there exists $n \geq x_m$ so that $\chi_n(i) = \chi^*(i)$ for all $i \leq n$, in particular for all $x_j \in X$. Since χ_n is an r-coloring of $[n]$, X is not monochromatic under χ_n and thus X is not monochromatic under χ.

Proof 2. Let T be the set of *all* functions $f: V \to [r]$. We topologize T by giving $[r]$ the discrete topology and giving T the induced function space topology. In other words, for all $v_1, \ldots, v_n \in V$, $\varepsilon_1, \ldots, \varepsilon_n \in [r]$,

$$S_{v_1, \ldots, v_n, \varepsilon_1, \ldots, \varepsilon_n} = \{f: f(v_i) = \varepsilon_i, 1 \leq i \leq n\}$$

(a "slice") is both open and closed, and these S form a basis for the topology. T is the direct product of "$|V|$ copies of $[r]$." The set $[r]$ is finite and hence forms a compact topological space. The *Tychonoff theorem* (and here we are using the Axiom of Choice) states that the product of compact spaces is compact. Hence T is compact.

For every *finite* $W \subseteq V$ let F_W denote the set of functions $f \in T$ so that no $X \in E$, $X \subseteq W$ is monochromatic. The set F_W consists of those functions that are r-colorations when restricted to W. Each F_W is closed (and open) since it is the union of a finite number of slices $S_{w_1, \ldots, w_n, \varepsilon_1, \ldots, \varepsilon_n}$ ($W = \{w_1, \ldots, w_n\}$). Each $F_W \neq \varnothing$ since, by assumption, there is an r-coloring of each finite set W. Clearly, if $W \subseteq W'$, $F_W \supseteq F_{W'}$. Applying this, we find that if $W_1, \ldots, W_m$ are finite subsets of V then

$$F_{W_1} \cap \cdots \cap F_{W_m} \supseteq F_{W_1 \cup \cdots \cup W_m}.$$

Now $W_1 \cup \cdots \cup W_m$, a finite union of finite sets, is finite so $F_{W_1 \cup \cdots \cup W_m} \neq \varnothing$. Thus $\{E_W: W \subseteq V, W \text{ finite}\}$ is a family of closed sets satisfying the finite intersection property: any finite intersection of the F_W is nonvoid. In a compact topological space, if a family of closed sets $\mathscr{F}$ satisfies the finite intersection property then $\cap \mathscr{F} \neq \varnothing$; that is, there exists $f: V \to [r]$, $f \in F_W$, for all $W \subseteq V, W$ finite. This f is the desired coloring, for if $X \in E$, so X is finite, $f \in F_X$, and therefore X is not monochromatic under f.

In most applications of the Compactness principle to Ramsey results, $V = N$ or $[N]^k$. We restate our theorem for these particular cases.

Compactness Principle (Version B). Let k be a fixed positive integer. Let $\mathscr{A}$ be a family of finite subsets of N. Suppose that, for any r-coloring of $[N]^k$, there is an $A \in \mathscr{A}$ so that $[A]^k$ is monochromatic. Then there exists

n_0 so that, for $n \geq n_0$, if $[n]^k$ is r-colored there is an $A \in \mathscr{A}$, $A \subseteq [n]$, so that $[A]^k$ is monochromatic.

Compactness Principle (Version C). Let k be a fixed finite positive integer. Let $\mathscr{A}$ be a family of finite subsets of N. Suppose that for any finite coloring of $[N]^k$ there is an $A \in \mathscr{A}$ such that $[A]^k$ is monochromatic. Then for all r, there exists $n_0(r)$ such that, for $n \geq n_0(r)$, if $[n]^k$ is r-colored there is an $A \in \mathscr{A}$, $A \subseteq [n]$, such that $[A]^k$ is monochromatic. Often technical details in the proof of a Ramsey theorem (e.g., just *how* large n has to be that . . .) vanish in the "infinite case."

Theorem 5. For any finite coloration χ of $[N]^2$ there exists $A \subseteq N$, A infinite, so that $[A]^2$ is monochromatic.

Proof. (Following Proof 2 of Ramsey's theorem abridged). Define, for all $i \in N$, infinite sets S_i and elements $x_i \in S_i$ as follows:

(i) $S_1 = N$.
(ii) Having chosen S_i, choose $x_i \in S_i$ arbitrarily.
(iii) Having selected $x_i \in S_i$, set

$$T_j = \{u \in S_i : \chi(x_i, u) = j\} .$$

The T_j give a finite partition of $S_i - \{x_i\}$, assumed infinite. Set S_{i+1} equal to one of the infinite T_j.

The sequence $x_1, x_2, \ldots$ has the property that, for $i < j, k,$

$$\chi(x_i, x_j) = \chi(x_i, x_k)$$

[since $x_j \in S_j \subseteq S_{i+1}$, $x_k \in S_k \subseteq S_{i+1}$, and $\chi(x_i, u)$ is constant over $u \in S_{i+1}$]. Induce a coloring χ^* of the singletons x_i: $\chi^*(x_i) = $ that color equal to $\chi(x_i, x_j)$ for all $j > i$. Now χ^* forms a finite partition of an infinite set so there is a color j and an infinite subsequence $X' = x_{i_1}, x_{i_2}, \ldots$ such that

$$\chi^*(x_{i_s}) = j \qquad \text{for all } s .$$

For any $1 \leq s < t,$

$$\chi(x_{i_s}, x_{i_t}) = \chi^*(x_{i_s}) = j$$

so $[X']^2$ is monochromatic.

Corollary 6. For all l, r there exists n_0 so that, for $n \geq n_0$,

$$n \to (l)_r .$$

Proof. Compactness principle (Version C).

It is interesting that proofs using the Compactness principle do not give any specific n_0 such that, for $n \geq n_0$, the Ramsey property holds. In actual practice we are usually able to replace the argument on N with an argument that works for all $n \geq n_0$ for some specific n_0. In Section 6.3 we discuss a situation where such a replacement is not possible in a certain logical sense.

Questions about extensions of Ramsey-type theorems to infinite sets are interesting per se. The subject of "Infinite Ramsey theory" has a long, interesting literature. We give some glimpses into the field in Section 6.4.

1.6 A BROADER PERSPECTIVE

H. Burkill and L. Mirsky state, "There are numerous theorems in mathematics which assert, crudely speaking, that every system of a certain class possesses a large subsystem with a higher degree of organization than the original system." The existence of the Ramsey number $n = R(k)$ is their first example. For any graph G on n vertices there is a large (size k) subsystem (subgraph) of a high degree of organization (either complete or independent). Their class of problems includes, for example, the Bolzano–Weierstrass theorem that every bounded sequence of complex numbers contains a convergent subsequence. This class is thus far broader than the Ramsey theory to which we are attempting to restrict our attention. We do wish to mention some of the result in this broader class that both are interesting in their own right and seem close in spirit to our Ramsey theory.

R. P. Dilworth proved that any partial order P on at least $ab + 1$ elements contains either a chain of length $a + 1$ or an antichain of size $b + 1$. Dilworth's theorem is of fundamental importance and clearly fits into the Burkill–Mirsky setting. Let us note that if $ab + 1$ is replaced by the Ramsey function $R(a + 1, b + 1)$ the result is a corollary of Ramsey's theorem. Given a partial order p, we color a pair $\{x, y\}$ red if x, y are comparable, and blue if they are not. A red $(a + 1)$-set yields the desired chain; a blue $(b + 1)$-set yields the desired antichain.

P. Erdös and G. Szekeres proved that any sequence of length $n^2 + 1$ contains a monotone subsequence of length $n + 1$. Again, with $n^2 + 1$

replaced by $R(n + 1, n + 1)$ this follows from Ramsey's theorem. Given a sequence $\{a_i\}$, we color $\{i, j\}_<$ red if $a_i < a_j$, and blue otherwise. The Infinite Ramsey theorem similarly implies that any infinite sequence contains an infinite monotonic subsequence.

P. Erdös and L. Moser proved that every tournament on n players contains a transitive subtournament on $v(n)$ players, where $v(n)$ is a function tending to infinity with n. Here a tournament is a directed graph on n points so that, for all distinct x, y either $(x, y) \in T$ or $(y, x) \in T$, but not both, and a tournament is transitive if there exists a total ordering $<$ such that $(x, y) \in T$ iff $x < y$. Again the existence of $v(n)$ follows from Ramsey's theorem, though the actual bounds achieved by Erdös and Moser are stronger.

1.7 ORIGINAL PAPERS: RAMSEY AND ERDÖS–SZEKERES

Frank Plumpton Ramsey was a remarkable man. He was a child of Cambridge (his father was president of Magdalene) and spent nearly all his life there. He worked in several areas, always with keen insight and intelligence. He died in 1930 as he was approaching his twenty-eighth birthday, at the height of his intellectual powers.

One of Ramsey's many interests was economics, and he was part of the Cambridge circle headed by J. M. Keynes. He wrote only two papers in the field, "A Contribution to the Theorem of Taxation" (March 1927, *The Economic Journal*) and "A Mathematical Theory of Savings" (December 1928, *ibid.*). Keynes said of the latter paper: "[It] is one of the most remarkable contributions to mathematical economics ever made, both in respect of the intrinsic importance and difficulty of its subject, the power and elegance of the technical methods employed, and the clear purity of illumination with which the writer's mind is felt by the reader to play about its subject." Indeed, Keynes's judgment has stood the test of time, and today Ramsey's work is widely quoted in mathematical economics literature.

Ramsey's main interests were philosophy and mathematical logic. He was deeply influenced by Russell and Whitehead's *Principia Mathematica* and proposed a Theory of Types with certain advantages over that used by Russell and Whitehead. He helped translate and was greatly interested in the work of Wittgenstein. G. E. Moore wrote:

> [Ramsey] combined very exceptional brilliance with very great soundness of judgment in philosophy. He was an extraordinarily clear thinker: no one could avoid more easily than he the sort of confusions of thought to which

even the best philosophers are liable, and he was capable of apprehending clearly, and observing consistently, the subtlest distinctions. . . . I always felt with regard to any subject which we discussed, that he understood it much better than I did, and where (as was often the case) he failed to convince me, I generally thought the probability was that he was right and I wrong and that my failure to agree with him was due to a lack of mental powers on my part.

One feels, reading commentary on Ramsey's philosophical work, that he was only beginning to make major contributions to the subject at the time of his death.

And now we come to Ramsey's theorem. His paper [Ramsey, 1930] is indeed "On a Problem of Formal Logic." Although he recognized that Ramsey's theorem had independent interest, he was mainly concerned with its application to logic. Perhaps his view speaks of a time when combinatorial analysis was still regarded as "bargain basement topology" by the mainstream of mathematical thought. Yet it seems eminently suitable that this branch of combinatorial analysis be graced with the name of Frank Plumpton Ramsey.

Ramsey begins his paper with the infinite version of Ramsey's theorem. We include below his original proof. Brevity was not an admirable trait in that era, and authors preferred a lengthy discussion to the terse Theorem–Proof–Corollary style of today. Despite the 50-year gap in notation the paper reads with remarkable clarity.

THEOREM A.   *Let Γ be an infinite class, and μ and r positive integers; and let all those sub-classes of Γ which have exactly r members, or, as we may say, let all r-combinations of the members of Γ be divided in any manner into μ mutually exclusive classes C_i ($i = 1, 2, \ldots, \mu$), so that every r-combination is a member of one and only one C_i; then, assuming the Axiom of Selections, Γ must contain an infinite sub-class Δ such that all the r-combinations of the members of Δ belong to the same C_i.*

Consider first the case $\mu = 2$. (If $\mu = 1$ there is nothing to prove.) The theorem is trivial when r is 1, and we prove it for all values of r by induction. Let us assume it, therefore, when $r = \rho - 1$ and deduce it for $r = \rho$, there being, since $\mu = 2$, only two classes C_i, namely C_1 and C_2.

It may happen that Γ contains a member x_1 and an infinite sub-class Γ_1, not including x_1, such that the ρ-combinations consisting of x_1 together with any $\rho - 1$ members of Γ_1, all belong to C_1. If so, Γ_1 may similarly contain a member x_2 and an infinite sub-class Γ_2, not including x_2, such that all the ρ-combinations consisting of x_2 together with $\rho - 1$ members of Γ_2, belong to C_1. And, again, Γ_2 may contain an x_3 and a Γ_3 with similar properties, and so on indefinitely. We thus have two possibilities: either we can select in this way two infinite sequences of members of $\Gamma(x_1, x_2, \ldots, x_n, \ldots)$, and of infinite sub-classes of $\Gamma(\Gamma_1, \Gamma_2, \ldots, \Gamma_n, \ldots)$, in which x_n is always a member of Γ_{n-1}, and Γ_n a sub-class of Γ_{n-1} not including x_n, such that all the ρ-combinations consisting of x_n, together with $\rho - 1$ members of Γ_n, belong to C_1; or else the process of selection will fail at a certain stage, say the n-th, because Γ_{n-1} (or if $n = 1$, Γ itself) will contain no member x_n and infinite sub-class Γ_n not including x_n such that all the ρ-combinations consisting of x_n together with $\rho - 1$ members of Γ_n belong to C_1. Let us take these possibilities in turn.

If the process goes on forever let Δ be the class $(x_1, x_2, \ldots, x_n, \ldots)$. Then all these x's are distinct, since if $r > s$, x_r is a member of Γ_{r-1} and so of $\Gamma_{r-2}, \Gamma_{r-3}, \ldots$, and ultimately of Γ_s which does not contain x_s. Hence Δ is infinite. Also all ρ-combinations of members of Δ belong to C_1; for if x_s is the term of such a combination with least suffix s, the other $\rho - 1$ terms of the combination belong to Γ_s, and so form with x_s a ρ-combination belonging of C_1. Γ therefore contains an infinite sub-class Δ of the required kind.

Suppose, on the other hand, that the process of selecting the x's and Γ's fails at the n-th stage, and let y_1 be any member of Γ_{n-1}. Then the $(\rho - 1)$-combinations of members of $\Gamma_{n-1} - (y_1)$ can be divided into two mutually exclusive classes C_1' and C_2' according as the ρ-combinations formed by adding to them y_1 belong to C_1 or C_2, and by our theorem (A), which we are assuming true when $r = \rho - 1$ (and $\mu = 2$), $\Gamma_{n-1} - (y_1)$ must contain an infinite sub-class Δ_1 such that all $(\rho - 1)$-combinations of the members of Δ_1 belong to the same C_i'; i.e. such that the ρ-combinations formed by joining y_1 to $\rho - 1$ members of Δ_1 all belong to the same C_i. Moreover, this C_i cannot be C_1, or y_1 and Δ_1 could be taken to be x_n and Γ_n, and our previous process of selection would not have failed at the n-th stage. Consequently the ρ-combinations formed by joining y_1 to $\rho - 1$ members of Δ_1 all belong to C_2. Consider now Δ_1 and let y_2 be any of its members. By repeating the preceding argument $\Delta_1 - (y_2)$ must contain an infinite sub-class Δ_2 such that all the ρ-combinations got by joining y_2 to $\rho - 1$ members of Δ_2 belong to the same C_i. And, again, this C_i cannot be C_1, or, since y_2 is a member and Δ_2 a sub-class of Δ_1 and so of Γ_{n-1} which includes Δ_1, y_2 and Δ_2 could have been chosen as x_n and Γ_n and the process of selecting these would not have failed at the n-th stage. Now let y_3 be any member of Δ_2; then $\Delta_2 - (y_3)$ must contain an infinite sub-class Δ_3 such that all ρ-combinations consisting of y_3 together with $\rho - 1$ members of Δ_3, belong to the same C_i, which, as before, cannot be C_1 and must be C_2. And by continuing in this way we shall evidently find two infinite sequences $y_1, y_2, \ldots, y_n, \ldots$ and $\Delta_1, \Delta_2, \ldots, \Delta_n, \ldots$ consisting respectively of members and sub-classes of Γ, and such that y_n is always a member of Δ_{n-1}, Δ_n a sub-class of Δ_{n-1} not including y_n, and all the ρ-combinations formed by joining y_n to $\rho - 1$ members of Δ_n belong to C_2; and if we denote by Δ the class $(y_1, y_2, \ldots, y_n, \ldots)$ we have, by a previous argument, that all p-combinations of members of Δ belong to C_2.

Hence, in either case, Γ contains an infinite sub-class Δ of the required kind, and Theorem A is proved for all values of r, provided that $\mu = 2$. For higher values of μ we prove it by induction; supposing it already established for $\mu = 2$ and $\mu = \nu - 1$, we deduce it for $\mu = \nu$.

The r-combinations of members of Γ are then divided into ν classes C_i $(i = 1, 2, \ldots, \nu)$. We define new classes C_i' for $i = 1, 2, \ldots, \nu - 1$ by

$$C_i' = C_i (i = 1, 2, \ldots, \nu - 2),$$
$$C_{\nu-1}' = C_{\nu-1} + C_\nu.$$

Then by the theorem for $\mu = \nu - 1$, Γ must contain an infinite sub-class Δ such that all r-combinations of the members of Δ belong to the same C_i'. If, in this C_i', $i \leqslant \nu - 2$, they all belong to the same C_i, which is the result to be proved; otherwise they all belong to $C_{\nu-1}'$, i.e. either to $C_{\nu-1}$ or to C_ν. In this case, by the theorem for $\mu = 2$, Δ must contain an infinite sub-class Δ' such that the r-combinations of members of Δ' either all belong to $C_{\nu-1}$ or all belong to C_ν; and our theorem is thus established.

Ramsey then proceeds to the finite analogue. He does not mention the possibility of a Compactness argument. We paraphrase his argument in modern language.

Theorem 7. $\forall r, n, k, n + k \geqslant r$, $\exists m_0$ so that, for $m \geqslant m_0$, if $[m]^r$ is 2-colored there exist $S, T \subseteq [m]$, $|S| = n$, $|T| = k$, $S \cap T = \varnothing$, so that all r-subsets of $S \cup T$ containing at least one $x \in S$ are the same color.

Proof. We shall define $m_0(r, n, k)$ for all r, n, k. We use induction on r. The case $r = 1$ is trivial; take $m_0 = \max(2n - 1, n + k)$. Assume the result for all $r' < r$.

For $n = 1$, all k, we prove we may take $m_0(r, 1, k) = 1 + m_0(r - 1, k, 0)$. Let $|U| = 1 + m_0(r - 1, k, 0)$ and fix a 2-coloring

$$\chi : [U]^r \to \{\text{red, blue}\}.$$

Select $x \in U$ arbitrarily, and define a 2-coloring χ' of $[U - \{x\}]^{r-1}$ by

$$\chi'(V) = \chi(V \cup \{x\}).$$

As $|U - \{x\}| = m_0(r - 1, k, 0)$, we find T, $|T| = k$ so that χ' is constant on $[T]^{r-1}$. The theorem follows for $n = 1$ by setting $S = \{x\}$.

Fix $r > 1$, $n > 1$, k. By induction assume that the theorem holds for

$r' < r$, all n, k and for $r' = r, n' < n$, and all k, Defined $F(k) = m_0(r, 1, x)$, and set $F^{(t)}$ equal to the tth iterate of F. $F^{(t)}$ is defined for all t. Now let

$$|U| > m_0(r, n - 1, F^{(n)}(\max(r - 1, k)))$$

and fix a 2-coloring

$$\chi: [U]^r \to \{\text{red, blue}\} .$$

We can find disjoint $S, T, |S| = n - 1, |T| = F^{(n)}(\max(r - 1, k))$ by the induction hypothesis so that χ is monochromatic, say red, on all $X \subseteq S \cup T$, $|X| = r$, $X \cap S \neq \emptyset$. We find $t_1 \in T$, $T_1 \subseteq T - \{t_1\}$, $|T_1| = F^{(n-1)}(\max(r - 1, k))$ so that all $X \subseteq \{t_1\} \cup T_1$, $|X| = r, t_1 \in X$ are monochromatic. If they are red then our theorem is satisfied by $S' = S \cup \{t_1\}$ and T' equal to any subset of T_1 of cardinality k. We assume that they are blue. Now we find $t_2, T_2, |T_2| = F^{(n-2)}(\max(r - 1, k))$, where all $X \subseteq \{t_2\} \cup T_2$, $|X| = r, t_2 \in X$ are monochromatic. Again we are finished if they are red so we assume that they are blue. We continue to find $t_1, \ldots, t_n, T_n$ with $|T_n| = \max(r - 1, k)$ so that if $U \subseteq \{t_1, \ldots, t_n\} \cup T_n$ and $U \cap \{t_1, \ldots, t_n\} \neq \emptyset$ then $\chi(U)$ is blue. The sets $\{t_1, \ldots, t_n\}$ and T_n (or, rather, any k-element subset of T_n) satisfy the induction hypothesis. This completes the proof.

This theorem with $k = 0$ gives Ramsey's theorem for two colors—a simple induction on the number of colors gives the full result.

Ramsey noted that application of this proof gives

$$R(n) \leq 2^{n(n-1)/2} ,$$

but he improves this to

$$R(n) \leq n!$$

He states, intriguingly, "But this value is, I think, still much too high." There is no evidence that he was aware either the exponential upper bound or the exponential lower bound.

Ramsey's original application, and purpose, for Ramsey's theorem is of interest in its own right. We rephrase it, placing the combinatorial character in the following two theorems.

Theorem 8. For all $n_1, \ldots, n_k, t$ there exists m' so that if $m > m'$ the following holds: Let $|S| = m$, and let $[S]^t$ be n_i-colored for $1 \leq i \leq k$.

Then there exists $T \subseteq S$, $|T| = t$ so that, for each i, $1 \leq i \leq k$, $[T]^i$ is monochromatic.

Proof. Define a sequence $m_1, \ldots, m_k$ inductively so that

$$m_1 \to (t)^1_{n_1},$$

$$m_i \to (m_{i-1})^i_n, \qquad 2 \leq i \leq k.$$

We prove we may take $m' = m_k$. Let $|S| = m > m_k$ and fix a coloring of $[S]^{\leq k}$. We find $S_{k-1} \subseteq S$, $|S_{k-1}| = m_{k-1}$ so that $[S_{k-1}]^k$ is monochromatic. We then find $S_{k-2} \subseteq S_{k-1}$, $|S_{k-2}| = m_{k-2}$ so that $[S_{k-2}]^{k-1}$ is mono-chromatic. Continuing, we find a sequence $S = S_k \supseteq S_{k-1} \supseteq \cdots \supseteq S_1 \supseteq S_0$, where $|S_0| = t$ and, for all $1 \leq i \leq k$, $[S_0]^i$ is monochromatic (since $S_0 \subseteq S_{i-1}$ so $[S_0]^i \subseteq [S_{i-1}]^i$).

It is surprising to find such a sophisticated use of Ramsey's theorem in the original paper of Ramsey. We now require a definition. All elements are considered integers (or, more generally, members of a set totally ordered by $<$).

DEFINITION. $(x_1, \ldots, x_k) \sim (y_1, \ldots, y_k)$ if for all i, j,

$$(x_i < x_j \text{ iff } y_i < y_j), \qquad (x_i = x_j \text{ iff } y_i = y_j), \qquad \text{and}$$

$$\times (x_i > x_j \text{ iff } y_i > y_j).$$

Then $\sim$ is clearly an equivalence relation. Intuitively it means "has the same ordering as."

DEFINITION. Let R be a k-ary relation. We say that R is *canonical* on a set S if

$$(x_1, \ldots, x_k) \sim (y_1, \ldots, y_k) \Rightarrow [R(x_1, \ldots, x_k) \Leftrightarrow R(y_1, \ldots, y_k)]$$

for all $x_1, \ldots, x_k, y_1, \ldots, y_k \in S$.

For example, there are exactly eight canonical binary relations: "false," ">," "=," "<," "$\leq$," "$\neq$," "$\geq$," "true."

Theorem 9. For all $b_1, \ldots, b_k, t$ there exists m' so that, for $m > m'$, the following holds: Let $\mathfrak{R}$ be a set of relations on $[m]$ consisting of b_i i-ary relations, $1 \leq i \leq k$. There exists S, $|S| = t$ on which all $R \in \mathfrak{R}$ are canonical.

Proof. We define an equivalence class on $[m]^i$ for $1 \le i \le t$. Let

$$X = \{x_1, \ldots, x_i\}_<, Y = \{y_1, \ldots, y_i\}_< \in [m]^i .$$

We say that $X \sim Y$ if, for every $j, i \le j \le k$, for every sequence $w_1, \ldots, w_j$ such that $\{w_1, \ldots, w_j\} = \{1, \ldots, i\}$ (though perhaps with repetitions), and for every j-ary $R \in \Re$,

$$R(x_{w_1}, \ldots, x_{w_j}) \Leftrightarrow R(y_{w_1}, \ldots, y_{w_j}) .$$

This gives a finite number of equivalence classes. For example, if $\Re$ consists solely of binary relations $R_1, \ldots, R_b$ then $\{x, y\}_<$ is "colored" by the truth values of $R_i(x, y)$ and $R_i(y, x)$ for $1 \le i \le b$. Hence there are exactly 2^{2b} possible equivalence classes. Singletons $\{x\}$ are colored by the truth values of $R_i(x, x)$ for $1 \le i \le b$, with 2^b possible equivalence classes.

By Theorem 8 we select m' in such a way that there exists S, $|S| = t$ so that, for $1 \le i \le k$, all $X \in [S]^i$ are in the same equivalence class—hence all $R \in \Re$ are canonical on S. This completes the proof.

Now we come to the application of these combinatorial theorems to mathematical logic.* Let Q be an axiom system in first-order logic involving Boolean expressions, equality, k-ary relations, and no existential quantifiers—that is, all statements are universally quantified. Here are two examples.

$$Q_1: \forall_x \forall_y (x = y) \veebar (xRy) \veebar (yR_x)$$

$$\forall_x \forall_y \forall_z \{xRy \wedge yRz\} \Rightarrow xRz$$

$$Q_2: \forall_x \forall_y x \ne y \Rightarrow [xRy \veebar xBy]$$

$$\forall_x \forall_y [xRy \Leftrightarrow yRx] \wedge [xBy \Leftrightarrow yBx]$$

$$\forall_x \forall_y \forall_z x \ne y \ne z \ne x \Rightarrow (\sim[xRy \wedge yRz \wedge xRz])$$

$$\wedge (\sim[xBy \wedge yBz \wedge xBz])$$

Let $t(Q)$ denote the number of variables used in Q: here $t(Q_1) = t(Q_2) = 3$ (i.e., x, y, z). A model $\mathfrak{m}$ of Q is called canonical if all relations in the model are canonical.

Theorem 10. For all $b_1, \ldots, b_k, t$ there exists m' so that for $m > m'$ the following holds: Let Q be an axiom system with b_i i-ary relations,

---

* These results require some familiarity with mathematical logic. They are not required for the remainder of the book.

$1 \le i \le k$, $t(Q) = t$. Then there exists a model $\mathfrak{m}$, $|\mathfrak{m}| = m$ iff there exists a canonical model $\mathfrak{m}'$, $|\mathfrak{m}'| = t$.

Proof. The m' is taken identically as in Theorem 9.

The "if" part is immediate as a canonical model $\mathfrak{m}'$ can be extended to any ordered set in the "canonical" way. As an illustration, Q_1 has canonical model $\mathfrak{m}' = \{1, 2, 3\}$ where R is $<$. For any m we define a model $\mathfrak{m}$ on $[m]$, R being $<$.

The "only if" part follows from Theorem 9. Assume the existence of $\mathfrak{m}$, $|\mathfrak{m}| = m$. There exists a subset S of cardinality t on which all relations are canonical so that the restriction of $\mathfrak{m}$ to S is a canonical model. Here, critically, there are no existential quantifiers so that the restriction of a model to a subset is still a model. Q_2 provides a good example as one can see quickly that there are no canonical models, and we trust that by this point the reader can demonstrate that there are no models on six or more elements.

Although Ramsey's theorem is accurately attributed to Frank Ramsey, its popularization stems from the classical 1935 paper of P. Erdös and G. Szekeres. Esther Klein (later to become Esther Szekeres) had discovered the following curious result: Given five points in a plane, some four form a convex quadrilateral. A generalization was conjectured: For all n there exists N such that for any N points in a plane there are n that form a convex n-gon. Szekeres, in a forward to the collected combinatorial works of Erdös, gives an account of the climate, social and mathematical, surrounding their discoveries. We quote from his account:

> I have no clear recollection how the generalization actually came about; in the paper we attributed it to Esther, but she assures me that Paul and much more to do with it. We soon realized that a simple-minded argument would not do and there was a feeling of excitement that a new type of geometric problem emerged from our circle which we were only too eager to solve. For me the fact that it came from Epszi [Paul's nickname for Esther, short for "epsilon"] added a strong incentive to be the first with a solution and after a few weeks I was able to confront Paul with a triumphant "E.P., open your wise mind." What I really found was Ramsey's Theorem, from which it easily followed that there exists a number $N < \infty$ such that out of N points in the plane it is possible to select n points which from a convex n-gon. Of course at that time none of us knew about Ramsey.

Here is Szekeres' argument in our notation. Select N so that

$$N \to (n, 5)^4 .$$

Now color a four-element subset red if it forms a convex quadrilateral, and blue otherwise. Since there are no blue 5-sets there must be a red n-set. But it is not difficult to prove that, if every four points from a convex quadrilateral, the n points must form a convex n-gon.

Recently, another proof along similar lines was given by M. Tarsy. Select N so that $N \to (n)^3$. Now let N points in the plane be given, and number them $1, \ldots, N$ arbitrarily. Color $\{i, j, k\}_<$ red if traveling from i to j to k to i is in a clockwise direction, and blue if counterclockwise (both if collinear). Then there are n points ordered so that every triple has the same orientation, from which it follows easily that the n points form a convex n-gon. Tarsy was at the time an Israeli student who had been given this problem in an examination. Fortunately, he had been absent from class when the relevant material was discussed and so was forced to rely on his own imagination.

The classic 1935 paper also includes the result that any sequence of length $n^2 + 1$ contains a monotone subsequence of length $n + 1$. Strictly speaking, this result was not germane to the original problem, but the method of proof generalized and gave the second proof for the existence of N.

Let $N(n)$ denote the *minimal* value of N so that, for any N points in a plane, there are n that form a convex n-gon. This second proof (which Szekeres attributes entirely to Erdös) yielded the upper bound

$$N(n) \leq \binom{2n-4}{n-2} + 1 .$$

One can show that $N(n) > 2^{n-1}$ by a direct construction. Erdös, Szekeres, and Klein believe that $N(n) = 2^{n-2} + 1$ is the correct value. This remains an open problem.

It is difficult to overestimate the effect of this paper. The rediscovery of Ramsey's theorem and that of the Monotone Subsequence theorem were each of fundamental importance. Together they opened a new era in combinatiorial analysis. Both Szekeres and Turán consider these results to have been a decisive stage in Erdös' combinatorial studies. And certainly a major share of the interest in Ramsey theory in this generation is due to its popularization by Erdös. (See the chapter in Mope [2013].)

REMARKS AND REFERENCES

Ramsey [1930], Skolem [1933], and Erdös, Szekeres [1935] are the basic early references for Ramsey's theorem. We have generally followed the proofs of Skolem [1933].

§4. Turán's theorem may be found in almost any textbook on Graph theorem. See Turán [1954] or Turán [1941] (but in Hungarian) for the original proof. Motzkin and Straus [1965] give a short proof. Erdös and Sós [1969] discuss the relationship between Ramsey theorems and density theorems.

§5. The Compactness principle has no single discoverer. See Erdös [1950], Rado [1949], and Gottschalk [1951] and the general discussion in DeBruijn and Erdös [1951]. The Compactness principle is often called the Rado Selection principle.

§6. The quotation is from Burkill and Mirsky [1973]. Dilworth [1950]. Erdös and Szekeres [1935], Erdös and Moser [1964] give cited results. Seidenberg [1959] gives a particularly elegant proof of the Erdös–Szekeres Monotone Subsequence theorem.

§7. Comments of Szekeres are quoted from Erdös [1973]. Erdös and Szekeres [1962] give an interesting follow-up of their original paper.

2

Progressions

"Complete disorder is impossible."

T. S. Motzkin

2.1 VAN DER WAERDEN'S THEOREM

In 1927 B. L. van der Waerden published a proof of the following unexpected result.

Theorem 1 (Van der Waerden's Theorem). If the positive integers are partitioned into two classes then at least one of the classes must contain arbitrarily long arithmetic progressions.

This result, conjectured by I. Schur several years earlier, has turned out to be the seed to which much of the development of Ramsey theory may be traced. We examine several proofs of this theorem of van der Waerden and see how it leads naturally to various generalizations. Van der Waerden's personal account of his discovery is a classic work on the psychology of problem solving. We attempt to introduce the basic ideas as they actually occurred according to this account.

Historical Note. I. Schur, working on the distribution of quadratic residues in Z_p, first conjectured the result proved by van der Waerden. Van der Waerden heard of the conjecture through Baudet, a student at Göttingen at the time, and has referred to his result as Baudet's Conjecture in the literature. A brief account of Schur's contribution is given by A. Brauer in the preface to I. Schur-Gesammelte Abhandlungen (Springer-Verlag, 1973).

There are two rather harmless looking modifications we make in the statement of van der Waerden's theorem, both of which have a major impact on the proof. First, for each k we allow only a *finite* initial segment of integers (depending on k) to be partitioned so that at least one class is forced to contain an arithmetic progression of k terms. This modification, attributed to O. Schreier, is equivalent to the original assertion by the Compactness principle. Second, we allow the sets of integers to be partitioned into r classes instead of just two. This idea was suggested by E. Artin and is crucial to all known proofs of van der Waerden's theorem. Thus modified the statement is as follows:

> For all positive integers k and r, there exists an integer $W(k, r)$ so that, if the set of integers $\{1, 2, \ldots, W(k, r)\}$ is partitioned into r classes, then at least one class contains a k-term arithmetic progression.

To motivate the proof of the general theorem, we first examine a few small cases. Of course, for $k = 2$ and *any* r, the result is immediate [in fact, we may take $W(2, r) = r + 1$]. Let us consider the case $k = 3$, $r = 2$. We claim that we can take $W(3, 2) = 325$. To see this, assume that integers $\{1, 2, \ldots, 325\} \equiv [1, 325]$ are arbitrarily partitioned into two classes. Divide them into 65 blocks of length 5, that is,

$$[1, 325] = [1, 5] \cup [6, 10] \cup \cdots \cup [321, 325],$$

which we can write symbolically as

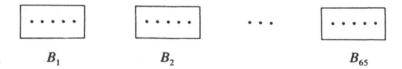

$$B_1 \qquad\qquad\qquad B_2 \qquad\qquad\qquad\qquad B_{65}$$

Since these integers are being split into $r = 2$ classes, that is, they are 2-colored, there are just $2^5 = 32$ possible ways to 2-color a block B_i. Thus, of the first 33 blocks B_i, some *pair* of blocks must be 2-colored in exactly the same way (by the Pigeon-Hole principle), say B_{11} and B_{26}. Look at this 2-coloring of $B_{11} = \{51, 52, 53, 54, 55\}$. Of the first *three* elements of B_{11}, that is, $\{51, 52, 53\}$, at least *two* of them must have the same color, say j and $j + d$. Since j and $j + d$ belong to $\{51, 52, 53\}$, $j + 2d$ belongs to B_{11}. (This is why we choose B_i to have length 5.) If $j + 2d$ has the same color as j (and $j + d$), we are done. Thus we may assume that it has the other color. A typical picture of the situation is shown in Fig. 2.1, where ● denotes red and ○ denotes blue. But now we are done, for if the integer $205 \in B_{41}$ is blue then $55, 130, 205$ is a blue arithmetic progression (AP), and if 205 is red then $51, 128, 205$ is a red AP.

What we have really done is to "focus" two two-term APs having different colors on the integer 205 so that, no matter *what* color it has, it must form the third term of *some* monochromatic AP.

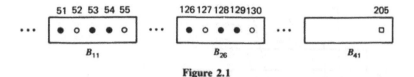

Figure 2.1

Let us use the same idea to find a value for $W(3, 3)$. This time, however, we start with an arbitrary 3-coloring of the first $7(2 \cdot 3^7 + 1)(2 \cdot 3^{7(2 \cdot 3^7 + 1)} + 1)$ integers! We first divide these integers into $2 \cdot 3^{7(2 \cdot 3^7 + 1)} + 1$ blocks B_i of $7(2 \cdot 3^7 + 1)$ each. Now, there are only $3^{7(2 \cdot 3^7 + 1)}$ different ways to 3-color each B_i so that, among the *first* $3^{7(2 \cdot 3^7 + 1)} + 1$ of them, at least two, say B_{i_1} and $B_{i_1 + d_1}$, have *exactly the same 3-colorings*. (The reason we use $2 \cdot 3^{7(2 \cdot 3^7 + 1)} + 1$ blocks is to ensure that the block $B_{i_1 + 2d_1}$ is well defined; we shall soon need to select an element from it.) Next, for each i, we divide the integers in B_i into $2 \cdot 3^7 + 1$ subblocks $B_{i,j}$ of 7 each. Since there are just 3^7 ways of 3-coloring each $B_{i,j}$, among the first $3^7 + 1$ blocks $B_{i_1, j}$, $1 \leq j \leq 3^7 + 1$, at least two, say B_{i_1, i_2} and $B_{i_1, i_2 + d_2}$, have exactly the same 3-colorings. Finally, in the *first* four elements of B_{i_1, i_2}, some color must occur twice; say that i_3 and $i_3 + d_3$ each are red. Since $i_3 + 2d_3$ is also in B_{i_1, i_2}, $i_3 + 2d_3$ must have some other color, say blue. The situation is shown in Fig. 2.2.

Consider the block $B_{i_1, i_2 + 2d_2}$. By the choice of i_2 and d_2, this is a subblock of B_{i_1}. Also, since B_{i_1, i_2} and $B_{i_1, i_2 + d_2}$ have the same 3-coloring, the integers $i_3 + 7d_2$ and $i_3 + d_3 + 7d_2$ must be red and the integer $i_3 + 2d_3 + 7d_2$ must be blue. Thus the corresponding element $i_3 + 2d_3 + 14d_2$ of $B_{i_1, i_2 + 2d_2}$ must be, say, yellow, not red nor blue, because of the arithmetic progressions $i_3 + 2d_3, i_3 + 2d_3 + 7d_2, i_3 + 2d_3 + 14d_2$ and $i_3, i_3 + d_3 + 7d_2, i_3 + 2d_3 + 14d_2$. Of course, since B_{i_1} and $B_{i_1 + d_1}$ have exactly the same 3-coloring, exactly the same color pattern occurs in $B_{i_1 + d_1}$; that is, the integers $i_3 + 7(2 \cdot 3^7 + 1)d_1, i_3 + d_3 + 7(2 \cdot 3^7 + 1)d_1, i_3 + 7d_2 + 7(2 \cdot 3^7 + 1)d_1$, and so on are red, the integers $i_3 + 2d_3 + 7(2 \cdot 3^7 + 1)d_1, i_3 + 2d_3 + 7d_2 + 7(2 \cdot 3^7 + 1)d_1$, and so on are blue, and the integers $i_3 + 2d_3 + 14d_2$ and $i_3 + 2d_3 + 14d_2 + 7(3^7 + 1)d_1$ are yellow.

Now consider the integer

$$m = i_3 + 2d_3 + 14d_2 + 14(3^7 + 1)d_1 \,.$$

There are three monochromatic two-term APs "focused" on m, each having a different color. The situation is shown in Fig. 2.3. If m is red then $i_3, i_3 + d_3 + 7d_2 + 7(3^7 + 1)d_1, m$ is a monochromatic AP. If m is

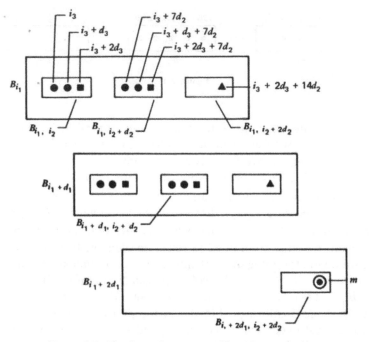

Figure 2.2 Forcing a three-term arithmetic progression.

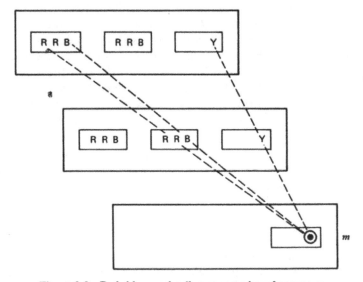

Figure 2.3 Red, blue, and yellow progressions focus on m.

blue then $i_3 + 2d_3$, $i_3 + 2d_3 + 7d_2 + 7(3^7 + 1)d_1$, m is a monochromatic AP. If m is yellow then $i_3 + 2d_2 + 14d_2$, $i_3 + 2d_3 + 14d_2 + 7(3^7 + 1)d_1$, m is a monochromatic AP. There are no other possibilities. We have shown that we may take $W(3, 3) = 7(2 \cdot 3^7 + 1)(2 \cdot 3^{7(2 \cdot 3^7 + 1)} + 1)$.

The proof of the general theorem is now just a double induction on k, the length of the progression desired, and r, the number of colors. Not only do we assume that $W(k, r - 1)$ exists, but we also assume that $W(k - 1, r')$ exists for *all* values of r'. We need the very large values of r' since, in general, we shall always divide the original set of integers into equal-sized blocks B_i of consecutive integers and *apply the induction hypothesis to the blocks*, which for our purposes behave in the same way that the integers do. If the integers are being r-colored then the blocks are $r^{|B_i|}$-colored (where $|B|$ denotes the cardinality of B). For this reason, the values we obtain for $W(k, r)$ are gigantic. (see Section 2.5.)

The one additional difficulty remaining to be overcome to complete the proof of van der Waerden's theorem along the lines just outlined is the choice of comprehensible notation. The interested reader will probably find it profitable at this point to complete this proof before going on.

A Short Proof. It is perhaps not surprising that by strengthening the hypothesis of van der Waerden's theorem, we obtain a somewhat stronger result that at the same time is a bit easier to prove. However, the basic structure of the proof is essentially the same as that of van der Waerden's original proof.

We define $m + 1$ *l-equivalence classes of* $[0, l]^m$. For $0 \leq i \leq m$ the set of $(x_1, \ldots, x_m) \in [0, l]^m$ in which l appears in the i rightmost positions and nowhere else forms an *l*-equivalence class. [For $i = 0$ this is all $(x_1, \ldots, x_m)$ in which l does not appear.] (Figure 2.4 shows the *l*-equivalence classes for $l = 4$ and $m = 2$.) The *l*-equivalence classes are disjoint. They partition a proper subset of $[0, l]m$; the remaining sequences are not used. For any $l, m \geq 1$ we define a statement $S(l, m)$:

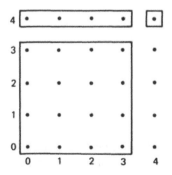

Figure 2.4 Critical equivalence classes.

For any r, there exists $N(l, m, r)$ so that for any function $C: [1, N(l, m, r)] \rightarrow [1, r]$ there exist positive integers $a, d_1, \ldots, d_m$ such that $C(a + \Sigma_{i=1}^m m_i d_i)$ is constant on each l-equivalence class of $[0, l]^m$.

The statement $S(l, 1)$ is equivalent to van der Waerden's theorem for l-term arithmetic progressions.

Theorem 2. $S(l, m)$ holds for all $l, m \geqslant 1$.

Proof.
  (i) $S(l, m) \Rightarrow S(l, m + 1)$.
      For a fixed r, let $M = N(l, m, r)$, $M' = N(l, 1, r^M)$ and suppose that $C: [1, MM'] \rightarrow [1, r]$ is given. Define $C': [1, M'] \rightarrow [1, r^M]$ so that $C'(k) = C'(k')$ iff $C(kM - j) = C(k'M - j)$ for all $0 \leqslant j < M$. By the inductive hypothesis, there exists a' and d' such that $C'(a' + xd')$ is constant for $x \in [0, l - 1]$. Let $I = [a'M - (M - 1), a'M]$. Since $S(l, m)$ can apply to the interval I then, by choice of M, there exist (renumbering for convenience) $a, d_2, \ldots, d_{m+1}$ with all sums $a + \Sigma_{i=2}^{m+1} x_i d_i$, $x_i \in [0, l]$, in I and with $C(a + \Sigma_{i=2}^{m+1} x_i d_i)$ constant on l-equivalence classes. Set $d_i' = d_i$ for $2 \leqslant i \leqslant m + 1$ and $d_1' = d'M$. Then $S(l, m + 1)$ holds.
  (ii) $S(l, m)$ for all $m \geqslant 1 \Rightarrow S(l + 1, 1)$.
      For a fixed r, let $C: [1, N(l, r, r)] \rightarrow [1, r]$ be given. Then there exists $a, d_1, \ldots, d_r$ such that, for $x_i \in [0, l]$, $a + \Sigma_{i=1}^r x_i d_i$ is bounded above by $N(l, r, r)$ and $C(a + \Sigma_{i=1}^r x_i d_i)$ is constant on l-equivalence classes. By the Pigeon-Hole principle there exist $1 \leqslant u < v \leqslant r + 1$ such that

$$C\left(a + \sum_{i=u}^r l d_i\right) = C\left(a + \sum_{i=v}^r l d_i\right).$$

Therefore

$$C\left(\left(a + \sum_{i=v}^r l d_i\right) + x\left(\sum_{i=u}^{v-1} d_i\right)\right)$$

is constant for $x \in [0, l]$. This proves $S(l + 1, 1)$.

2.2   THE HALES–JEWETT THEOREM

In its essence, van der Waerden's theorem should be regarded, not as a result dealing with integers, but rather as a theorem about finite sequences formed from finite sets. The Hales–Jewett theorem strips van der Waerden's theorem of its unessential elements and reveals the heart of Ramsey theory. It provides a focal point from which many results can be derived and acts as a cornerstone for much of the more advanced work. Without this result, *Ramsey theory* would more properly be called *Ramseyian theorems*.

We begin with notation. We define C_t^n, the n-cube over t elements, by

$$C_t^n = \{(x_1, \ldots, x_n): x_i \in \{0, 1, \ldots, t-1\}\}.$$

By a *line* in C_t^n we mean a set of (suitably ordered) points $\mathbf{x}_0, \ldots, \mathbf{x}_{t-1}$, $\mathbf{x}_i = (x_{i1}, \ldots, x_{in})$ so that in each coordinate $j, 1 \le j \le n$, either

$$x_{0j} = x_{1j} = \cdots = x_{t-1,j}$$

or

$$x_{sj} = s \qquad \text{for } 0 \le s < t,$$

and the latter occurs for at least one j (otherwise the $\mathbf{x}_i$ would be constant). For example, with $t = 4$, $n = 3$, $\{020, 121, 222, 323\}$ forms a line, as does $\{031, 131, 231, 331\}$. (In examples, parentheses and commas may be removed for clarity.)

Our definition differs from the ordinary geometric definition as, for example $\{02, 11, 20\}$ is not a line in C_3^2. The *reason* for this is that the cube is mean to be independent of the underlying set $\{0, 1, \ldots, t-1\}$. In other words, for any set $A = \{a_1, \ldots, a_t\}$ we may define

$$C_t^n = \{(x_1, \ldots, x_n): x_i \in A\}$$

and lines of C_t^n as those $\mathbf{x}_0, \ldots, \mathbf{x}_{t-1}$ so that in each coordinate j either the x_{ij} are constant or $x_{ij} = a_i$. All such cubes are combinatorially isomorphic. In this section we shall write our underlying set as $\{0, 1, \ldots, t-1\}$ solely to facilitate the exposition.

For $1 \le k \le n$ we define what we mean by a k-dimensional subspace of C_t^n. Let $\{1, \ldots, n\} = B_0 + B_1 + \cdots + B_k$, where $B_i \ne \varnothing$ for $1 \le i \le k$. (B_0 may be null.) Let

$$f: B_0 \to \{0, 1, \ldots, t-1\}$$

by any function. We define a map $\hat{f}: C_t^k \to C_t^n$ by

$$\hat{f}(y_1, \ldots, y_k) = (x_1, \ldots, x_n),$$

where

$$x_i = f(i) \qquad \text{for } i \in B_0,$$
$$x_i = y_j \qquad \text{for } i \in B_j.$$

A k-dimensional subspace is defined as a set that is the range of $\hat{f}$ for some choice of $B_0, B_1, \ldots, B_k, f$.

The real meaning of "k-dimensional subspace" may be gleaned from the following example, where $t = 3$, $n = 7$, $k = 2$, $B_1 = \{1, 2\}$, $B_2 = \{3, 4, 5\}$, $B_0 = \{6, 7\}$, $f(6) = 2$, $f(7) = 0$. The range of $\hat{f}$ is given by

$$
\begin{array}{ccc}
00\,000\,20 & 11\,000\,20 & 22\,000\,20 \\
00\,111\,20 & 11\,111\,20 & 22\,111\,20 \\
00\,222\,20 & 11\,222\,20 & 22\,222\,20
\end{array}
$$

The concept of k-dimensional subspace is clearly independent of the underlying set A. A line is a one-dimensional subspace.

A k-dimensional subspace S of C_t^n with underlying partition $B_0, B_1, \ldots, B_k$ in some fixed order is canonically isomorphic to C_t^k. In the example given above

$$\varphi: S \to C_3^2$$

given by

$$\varphi: (aabbb20) = ab,$$

is the isomorphism.

In all of our work there will be a clear ordering of the dimensions $B_0, B_1, \ldots, B_k$. Technically, we should refer to ordered k-dimensional subspaces. This will be tacitly assumed throughout.

Now we are in a position to state our fundamental result.

Theorem 3 (Hales–Jewett Theorem). For all r, t there exists $N' = HJ(r, t)$ so that, for $N \geqslant N'$, the following holds: If the vertices of C_t^N are r-colored there exists a monochromatic line.

We begin our proof with the equivalence classes of Section 2.1. For $0 \leqslant i \leqslant n$ the set of $(x_1, \ldots, x_n) \in C_{t+1}^n$ in which t appears in the i rightmost positions and nowhere else forms an equivalence class. We call a coloring of C_{t+1}^n *layered* if it is constant on all equivalence classes. (In a layered C_5^2 the enclosed sets of Fig. 2.4 are monochromatic. Also, a layered C_3^3 looks much like Fig. 2.2, minus any unessentials.) An (ordered) k-dimensional subspace is layered iff the coloration is layered when the subspace is identified canonically with C_{t+1}^k. (In the preceding example, 00 000 20, 00 111 20, 11 000 20, 11 111 20 would be the same color, and 00 222 20, 11 222 20 the same color.) A line is layered iff the first t points are monochromatic. When we say that a space is layered we always tacitly assume that it has a given coloration.

Example. Set $t = 27$, and set the underlying set $\mathscr{A}$ equal to the 26-letter English alphabet $A, B, \ldots, Z$ and ■ (space). The elements of C_{27}^n are then strings of length n. A string is left justified if all the spaces appear in the rightmost positions. In a layered coloring any two left-justified strings with the same number of letters have the same color. The line $\{\alpha A\alpha : \alpha \in \mathscr{A}\}$ is layered if $AAA, BAB, \ldots, ZAZ$ are the same color (with no restriction on the color of ■A■). If the two-dimensional space $\{\alpha A\beta\beta : \alpha, \beta \in \mathscr{A}\}$ is layered then $BALL, MASS$, and $PARR$ are the same color and MA■■, PA■■, and LA■■ are the same color.

We define two statements dependent on t, the cardinality of the underlying set:

$HJ(t)$: For all r there exists $N' = HJ(r, t)$ so that, for $N \geq N'$, if C_t^N is r-colored there exists a monochromatic line.

$LHJ(t)$: For all r, k there exists $M' = LHJ(r, t, k)$ so that, for $M \geq M'$, if C_{t+1}^M is r-colored there exists a layered k-dimensional subspace.

Our proof is by induction on t. We shall show that

$$HJ(t) \Rightarrow LHJ(t) \qquad \text{(Theorem 4)},$$
$$LHJ(t) \Rightarrow HJ(t+1) \qquad \text{(Corollary 6)}.$$

Proof of HJ(2). Set $HJ(r, 2) = r$. Consider the $N + 1$ points of C_2^N formed by a (possibly void) sequence of 1's followed by a (possibly void) sequence of 0's (e.g., for $N = 3$, the points 000, 100, 110, 111). For $N \geq r$ some two of these points must be the same color, and they form a monochromatic line. (Remember the definition—not every two points form a line!)

Technically, the induction argument we give can start at $t = 1$, and the reader might check that the inductive proof of $HJ(2)$ is essentially the proof we have given. However, as C_1^n is practically pointless (joke), the arguments $HJ(2) \Rightarrow LHJ(2) \Rightarrow HJ(3)$ might give fuller understanding.

Theorem 4. $HJ(t) \Rightarrow LHJ(t)$.

Proof. Assume $HJ(t)$. We prove $LHJ(t)$ by induction on k. As in van der Waerden's theorem, we prove $LHJ(t)$ for given k for all r simultaneously.

$k = 1$. Let $M' = HJ(r, t)$. Let $M \geqslant M'$, and r-color C_{t+1}^M. Inside C_{t+1}^M lies C_t^M, those points without coordinate value t. There is a monochromatic line in C_t^M that is a layered line in C_{t+1}^M.

$k \Rightarrow k + 1$. Here is the heart of the proof. Use the Induced Color method. Let $m = LHJ(r, t, k)$. Let $s = r^{(t+1)^m}$, the number of r-colorations of C_{t+1}^m. Set $m' = LHJ(s, t, 1)$, that is, $HJ(s, t)$. (Here m may be gigantic, but m' is unbelievably larger!) Take $LHJ(r, t, k+1) = m' + m$. Let $C_{t+1}^{m'+m}$ be r-colored by χ: $C_{t+1}^{m'+m} = C_{t+1}^{m'} \times C_{t+1}^m$ in a natural way. For $x \in C_{t+1}^{m'}$ and $y \in C_{t+1}^m$ write xy for their concatenation [e.g., $(2, 7, 5)(3, 6) = (2, 7, 5, 3\,6)$]. Define a coloring χ^* on $C_{t+1}^{m'}$, coloring $x \in C_{t+1}^{m'}$ by the color of xy for all $y \in C_{t+1}^m$. Formally

$$\chi^*(x) = \chi^*(x') \quad \text{iff} \quad \chi(xy) = \chi(x'y) \qquad \text{for all } y \in C_{t+1}^m .$$

As there are only(!) s colors, there exists a layered line $x_0, x_1, \ldots, x_{t-1}, x_t \in C_{t+1}^{m'}$ under χ^*. Now color C_{t+1}^m by

$$\chi^{**}(y) = \chi(x_i y) \qquad (0 \leqslant i \leqslant t - 1, \text{ as constant for those } i).$$

By induction there is a layered k-dimensional subspace $S \subseteq C_{t+1}^m$ under χ^{**}. Let

$$T = \{x_i s : 0 \leqslant i \leqslant t, s \in S\} \subseteq C_{t+1}^{m'+m} .$$

Let S have equivalence classes $S_0, S_1, \ldots, S_k$. Then T has equivalence classes

$$T_j = \{x_i s : 0 \leqslant i < t, s \in S_j\} , \qquad 0 \leqslant j \leqslant k ,$$

together with a T_{k+1} consisting of a single point beginning with x_t. Let $x_i s, x_{i'} s' \in T_j$ with $0 \leqslant j \leqslant k$. Then

$$\chi(x_i s) = \chi^{**}(s) = \chi^{**}(s') = \chi(x_{i'} s') .$$

The middle equality holds because s, s' are equivalent under χ^{**}, and the other equalities hold by the definition of χ^{**}. Hence T is a layered $(k+1)$-dimensional space, completing the induction.

Some intuition, going down instead of up, is useful here. Select $M = CH(r, t, k)$ so large that it may be written as $m_1' + m_1$, where m_1 is gigantic and m_1' is much, much bigger. An r-coloring of $C_{t+1}^M = C_{t+1}^{m_1'} \times C_{t+1}^{m_1}$ induces an s-coloring of $C_{t+1}^{m_1}$ ($s \gg m_1$ but $s \ll m_1'$) for which there exists a layered line $L_1 = \{x_0^{(1)}, \ldots, x_t^{(1)}\}$. On $L_1 \times C_{t+1}^{m_1}$ the color of $x_i^{(1)} y$ is independent of i if $i \neq t$. Color $y \in C_{t+1}^{m_1}$ by the color of $x_i^{(1)} y$, $i \neq t$. As m_1 is gigantic, write $m_1 = m_2' + m_2$, where $m_2' \gg m_2$ but m_2 is still gigantic. The r-coloring of $y \in C_{t+1}^{m_1} = C_{t+1}^{m_2'} \times C_{t+1}^{m_2}$ induces an s-coloring of $C_{t+1}^{m_2}$ for which there exists a layered line $L_2 = \{x_0^{(2)}, \ldots, x_t^{(2)}\}$. On $L_1 \times L_2 \times C_{t+1}^{m_2}$ the color of $x_i^{(1)} x_j^{(2)} y$ is independent of i if $i \neq t$ and is independent of both i and j if $i \neq t$ and $j \neq t$. Continue the entire procedure k times.

Theorem 5. A layered k-dimensional space with at most k colors contains a monochromatic line.

Proof. All ordered k-dimensional spaces over $t+1$ elements are combinatorially isomorphic. Hence it is sufficient to prove this result for C_{t+1}^k. This result corresponds to the focusing of progressions in van der Waerden's theorem. Pictorially this theorem is obvious; note in Fig. 2.4 that a layered 2-coloration of C_4^2 yields a monochromatic line. More formally, let C_{t+1}^k be a layered space and consider the special points $x_i, 0 \leq i \leq k$, defined by

$$x_i = (x_{i1}, \ldots, x_{ik}), \qquad x_{ij} = \begin{cases} t & \text{if } j \leq i, \\ 0 & \text{if } j > i. \end{cases}$$

(In C_4^2: 00, 40, 44.) By the Pigeon-Hole principle for some $u < v$, x_u and x_v are the same color, say red. Then the line $y_0, \ldots, y_t$, given by

$$y_s = (y_{s1}, \ldots, y_{sk}), \qquad y_{si} = \begin{cases} t & \text{if } i \leq u, \\ s & \text{if } u < i \leq v, \\ 0 & \text{if } u < i, \end{cases}$$

is red. (In C_4^2, if 00 and 44 are red the line 00, 11, 22, 33, 44 is red.)

Corollary 6. $LHJ(t) \Rightarrow HJ(t+1)$.

Proof. Given r, pick N' so that, for $N \geq N'$, and r-coloring of C_{t+1}^N contains a layered r-dimensional subspace. By Theorem 5 there must be a monochromatic line.

This completes the inductive proof of the Hales–Jewett theorem.

2.3 EXTENSIONS AND IMPLICATIONS

A simple trick gives a natural extension of the Hales–Jewett theorem.

Theorem 7 (Extended Hales–Jewett Theorem).   For all n, t, r there exists N' so that, for $N \geq N'$, the following holds: If the points of C_t^N are r-colored there exists a monochromatic n-dimensional subspace.

Proof.   We identity C_t^{ns} with $C_{t^n}^s$. The underlying t^n-set is C_t^n. We break $(x_1, \ldots, x_{ns}) \in C_t^{ns}$ into consecutive blocks of length n. Each block becomes a single coordinate, thus giving a set bijection between the two objects. A line in $C_{t^n}^s$ is identified, under the bijection, with an n-dimensional subspace of C_t^{ns}. (For example, the line 00, 11, 22, 33 in C_4^2 is identified with the two-dimensional space 0000, 0101, 1010, 1111 in C_2^4.) We set $s = HJ(r, t^n)$ and take $N' = ns$. An r-coloring of C_t^{ns} is identified with an r-coloring in $C_{t^n}^s$ that, by definition of s, has a monochromatic line. This line is identified with an n-dimensional subspace of C_t^{ns}, monochromatic under the original coloring. For $N \geq N'$, C_t^N contains $C_t^{N'}$ so the monochromatic subspace still exists.

Van der Waerden's theorem may be obtained as a corollary of the Hales–Jewett theorem. We identify the integers $a, 0 \leq a < t^N$, with the N-tuples $(a_1, \ldots, a_N)$ formed from the base-t representation $a = \sum_{i=1}^{N} a_i t^{i-1}$, $0 \leq a_i < t$. An r-coloring of $\{0, 1, \ldots, t^N - 1\}$ induces an r-coloring of C_t^N in which, for N sufficiently large, there is a monochromatic line that, in turn, translates back to a monochromatic AP of length t.

Let $V = \{v_0, \ldots, v_{t-1}\}$ be a subset of R^m. We say that $W = \{w_0, \ldots, w_{t-1}\}$ is *homothetic* to V if, under suitable ordering of W, there exist $c \in R$, $c \neq 0$, and $b \in R^m$ so that

$$w_i = cv_i + b, \qquad 0 \leq i < t.$$

In this case we may also write $W = cV + b$. In geometric terms, homothetic means "similar without rotating."

Theorem 8 (Gallai's Theorem).   Let the vertices of R^m be finitely colored. For all finite $V \subseteq R^m$ there exists a monochromatic W homothetic to V.

Proof.   The method of van der Waerden's theorem may be used to prove Gallai's theorem. It is simpler, however, to derive the latter as a corollary of the Hales–Jewett theorem. Fix the number of colors r and the set V, $|V| = t$. Let $N = HJ(r, t)$. Consider C_t^N to have underlying set V, so that the elements are sequences $(x_1, \ldots, x_N)$, $x_i \in V$. Define a map

$$\Psi: C_t^N \to R^m$$

by

$$\Psi(x_1, \ldots, x_N) = \sum_{i=1}^{N} k_i x_i$$

for real constants $k_1, \ldots, k_N$. Assume that Ψ is injective. An r-coloring of R^m [actually, of range (Ψ)] induces an r-coloring of C_t^N for which there is a monochromatic line that corresponds to a monochromatic $W \subset R^m$ homothetic to V.

We require Ψ to be injective, as otherwise a line in C_t^N could correspond to a single point in R^m. To achieve injectivity we appropriately choose $\{k_i\}$. For every $(x_1, \ldots, x_N) \neq (x_1', \ldots, x_N')$, both in C_t^N, we must have

$$\sum_{i=1}^{N} k_i(x_i - x_i') \neq 0.$$

The $\{k_i\}$ must be chosen to avoid only a finite set of equalities. Almost all choices of $\{k_i\}$ will suffice.

We may actually show a slightly stronger result. By selecting $k_i \in N$ we can assure a monochromatic $W = cV + b$, where $c \in N$.

An important example of Gallai's theorem is $V = \{(i, j): 0 \leq i, j < t\}$. We may then show the following: If N^2 is finitely colored there exist x_0, y_0, d so that all t^2 points of the form $(x_0 + id, y_0 + jd)$, $0 \leq i, j < t$, are the same color. Gallai's theorem may be considered a generalization of van der Waerden's theorem to higher dimensions.

A. W. Hales and R. I. Jewett, in their original paper, considered generalizations of tic-tac-toe, the classic children's game. In the original game (Fig. 2.5) two players alternately choose distinct elements of C_3^2, and a player wins if he or she has chosen an entire line (under the broader geometric definition). This game is well known to be a draw when properly played. Very recently the corresponding game for C_3^4 (again with the geometric definition of a line) has finally been resolved (the first player can always win). However, the winning strategy is extremely complicated. For all r, t, if N is sufficiently large the r-person "N-dimensional tic-tac-toe t-in a row" cannot end in an draw, even under

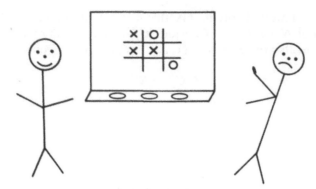

Figure 2.5 C_3^2.

the more restrictive definiton of a line. For $r = 2$ we may say more: for N sufficiently large the first player has a winning strategy. This is a standard Game theory argument: In a finite two-person, perfect information game with no draws, one of the two players must have a winning strategy. But in this game it cannot hurt to play first so that someone must be the first player! Let $GHJ(t)$ be the minimal N' so that, for $N > N'$, the first player wins tic-tac-toe on C_t^N. Then $GHJ(t) \leqslant HJ(2, t)$. However, the inequality need not be sharp; in fact, the exact nature of GHJ remains a puzzle.

2.4 SPACES—AFFINE AND VECTOR

Let A be an arbitrary, but fixed, finite field. We regard A^n as an n-dimensional space over A. We say that $X \subset A^n$ is a t-space if X is a t-dimensional affine subspace of A^n (i.e., a translate of a vector subspace of dimension t). The singleton sets are called 0-spaces. For this section only, let $[V]^t$ denote the class of t-spaces $T \subset V$. Our object is the following result.

Theorem 9 (Affine Ramsey Theorem). For all $r, t, , k$ there exists $n = N^{(t)}(k: r)$ so that if the t-spaces of A^n are r-colored there exists a k-space all of whose t-spaces have the same color.

Let $\dim(B) = u + 1$ and $p: B \to A^u$ be a surjective projection. Let $T \in [B]^t$. Then $p(T)$ has dimension either t or $t - 1$. If $\dim(p(T)) = t$ we call T transverse (relative to p). If $\dim(p(T)) = t - 1$ then $T = p^{-1}(p(T))$, and we call T vertical. Intuitively, p defines a vertical direction on B.

DEFINITION. A coloring $\chi: [B]' \to [r]$ is called special (relative to χ, p) if the color of a transverse t-space is determined by its projection. More formally $p(T_1) = p(t_2) \Rightarrow \chi(T_1) = \chi(T_2)$.

Lemma. For all u, r, t there exists $m = M^{(t)}(u: r)$ with the following property. Fix $p: A^{u+m} \to A^u$, the projection onto the first u coordinates. For any coloring $\chi: [A^{u+m}]' \to [r]$ there exists a $(u + 1)$-space B special relative to χ, p.

Proof. Let F_u denote the family of u-variable affine linear functions $f(x_1, \ldots, x_u) = c_0 + c_1 x_1 + \cdots + c_u x_u$, $c_0, \ldots, c_u \in A$ (possibly 0). We prove the lemma for

$$m = HJ(|F_u|, r^v),$$

where v is the number of t-subspaces of a u-space and HJ is the Hales–Jewett function. Fix $\chi: [A^{u+m}]' \to [r]$.
   Let $\tilde{f} = (f_1, \ldots, f_m)$, $f_i \in F_u$. We define a lifting

$$\tilde{f}: A^u \to A^{u+m}$$

by

$$\tilde{f}(x_1, \ldots, x_u) = (x_1, \ldots, x_u, y_1, \ldots, y_m), \qquad y_i = f_i(x_1, \ldots, x_u).$$

Clearly, $\tilde{f}$ is injective, linear, and inverse to p. We define (and this is the critical step) a coloring χ' on $(F_u)^m$ by

$$\chi'(\tilde{f}) = \chi'(\tilde{g}) \qquad \text{iff for all } T \in [A^u]', \chi(\tilde{f}(T)) = \chi(\tilde{g}(T)).$$

In other words, we color the lifting $\tilde{f}$ by the coloring of the range $\tilde{f}(A^u)$ of the lift. To be excessively formal we may define $\chi'(\tilde{f})$ as the function with domain $[A^u]'$ given by

$$(\chi'(\tilde{f}))(T) = \chi(\tilde{f}(T)).$$

   As χ' is an r^v-coloring there exists (and this is the central use of the Hales–Jewett theorem) in $(F_u)^m$ a "line" L monochromatic under χ'. By renumbering coordinates we may write

$$L = \{(f, \ldots, f, f_{s+1}, \ldots, f_m): f \in F_u\},$$

where $f_{s+1}, \ldots, f_m$ are fixed. We set

$$B = \bigcup_{\tilde{f} \in L} \tilde{f}(A^u)$$

$$= \{(x_1, \ldots, x_u, y_1, \ldots, y_m): y_i = y_1, 2 \le i \le s, y_i = f_i$$

$$\times (x_1, \ldots, x_u), s < i \le m\} .$$

B is the desired $(u + 1)$-space. Any transverse t-space $T \subset A^{u+m}$ may, by elementary linear algebra, by written as $T = \tilde{g}(p(T))$ for some $\tilde{g} = (g_1, \ldots, g_m) \in (F_u)^m$. When $T \subset B$, $T = \tilde{g}(p(T)) = \tilde{f}(p(T))$, where $f = (g_1, \ldots, g_1, f_{s+1}, \ldots, f_m) \in L$. Let $T_1, T_2 \in [B]^t$, $p(T_1) = p(T_2) = T \in [A^u]^t$. Then $T_1 = \tilde{f}_1(T)$, $T_2 = \tilde{f}_2(T)$, $\tilde{f}_1, \tilde{f}_2 \in L$. Hence

$$\chi(T_1) = \chi(\tilde{f}_1(T)) = \chi(\tilde{f}_2(T)) = \chi(T_2) .$$

Now the Affine Ramsey theorem is proved by a straightforward "induction on everything." We prove a strengthened result, as follows: For all $t, k_1, \ldots, k_r$ there exists $n = N^{(t)}(k_1, \ldots, k_r)$ so that if the t-spaces of A^n are r-colored there exists, for some $1 \le i \le r$, a k_i-space all of whose t-spaces are colored i.

The proof is a double induction, first on t [for all $(k_1, \ldots, k_r)$] and then on $(k_1, \ldots, k_r)$. For $t = 0$ the Affine Ramsey theorem follows directly from the Extended Hales–Jewett theorem. Assume the existence of n for $t' < t$ [all $(k_1, \ldots, k_r)$] and $t' = t$, $(k_1', \ldots, k_r') < (k_1, \ldots, k_r)$. We set

$$s = \max_{1 \le i \le r} N^{(t)}(k_1, \ldots, k_i - 1, \ldots, k_r) ,$$

$$u = N^{(t-1)}(s: r) ,$$

$$m = M^{(t)}(u: r) ,$$

$$n = u + m .$$

Let the t-spaces of A^n be r-colored arbitrarily by χ. By the definition of m (i.e., by the lemma) there is a $(u + 1)$-space B that is special under a projection $p: B \to A^u$. Induce a coloring χ' of $[A^u]^{t-1}$ by $\chi'(T) = \chi(p^{-1}(T))$. By the definition of u (i.e., induction on t) there exists an s-space $S \subset A^u$ monochromatic, say color 1, under χ'. Then $p^{-1}(S)$ is a special $(s + 1)$-space all of whose vertical t-spaces are color 1. Define a coloring χ'' of $[S]^t$ by

$$\chi''(T) = \chi(T') , \qquad \text{where } p(T') = T .$$

This is well defined since S is special, (i.e., project χ onto S). As $s \geqslant N^{(t)}(k_1 - 1, k_2, \ldots, k_r)$ [i.e., induction on $(k_1, \ldots, k_r)$], there exists $W' \subseteq S$ so that either

(i) $\dim(W') = k_1 - 1$; W' is color 1 under χ'',

or

(ii) $2 \leqslant i \leqslant r, \dim(W') = k_i$; W' is color i under χ''.

In case (ii), by linear algebra there exists a k-space $W \subseteq p^{-1}(W')$ so that $p(W) = W'$ (i.e., we may lift W' to W). Then W is color i under χ. In case (i) (the moment of induction) we set $W = p^{-1}(W')$. W is a vertical k_1-space of B. Let $T \in [W]^t$. If T is transverse, $\chi(T) = \chi''(p(T)) = 1$. If T is vertical $\chi(T) = \chi'(p(T)) = 1$. This completes the proof.

Corollary 10 (Vector Space Ramsey Theorem). For all $r, t, k \geqslant 1$ there exists $n = N^{(t)}(k: r)$ so that if the t-dimensional vector spaces of A^n are r-colored there exists a k-dimensional vector space all of whose t-dimensional vector subspaces have the same color.

Proof. Choose n to satisfy the Affine Ramsey theorem. A coloring χ of vector spaces T induces a coloring χ' of affine spaces by $\chi'(T + v) = \chi(T)$. (Every affine space T' may be written as $T' = T + v$, where T is a uniquely determined vector space.) There exists an affine k-space $W' = W + v$ monochromatic under χ'. Then W is monochromatic under χ.

The Vector Space Ramsey theorem was first conjectured by G.-C. Rota. One may view (see Section 1.3) Ramsey's theorem as a statement about the lattice of subsets of a set. The Vector Space Ramsey theorem is then the analogous statement for the lattice of subspaces of a vector space over a fixed finite field. This result was first shown by R. L. Graham, K. Leeb and B. L. Rothschild. The proof given is a simplified version due to J. Spencer.

2.5 ROTH'S THEOREM AND SZEMERÉDI'S THEOREM

The theorem of van der Waerden, while asserting the *existence* of a color class that contains arbitrarily long arithmetic progressions, does not specify *which* class is the appropriate one. In 1936, P. Erdös and P. Turán proposed the following conjecture:

If A is a set of positive integers with positive upper density, that is, satisfying

(*) $$\limsup_N \frac{|A \cap [1, N]|}{N} > 0,$$

then A contains arbitrarily long APs.

Thus this conjecture would imply that long APs always occur in the "most frequently occurring" color. In 1952, K. F. Roth proved that (*) implies that A must always contain at least a three-term AP. It was not until 1969 that E. Szemerédi showed that, in fact, (*) implies that A contains a four-term AP. In 1974, Szemerédi, in a masterpiece of combinatorial reasoning, settled the general conjecture affirmatively. In 1977, H. Furstenberg gave another proof, using methods of ergodic theory, of the Erdös–Turán conjecture. In Section 6.1 we discuss the relationships between the Szemerédi and Furstenberg proofs. However, both results are beyond the scope of this book.

In this section we prove Roth's theorem twice. We first give a combinatorial proof, due to Szemerédi, which contains many of the essentials of his general result. We follow this with Roth's original proof (slightly modified). This proof is one of the gems of Analytic Number theory, and the contrast with Szemerédi's proof is quite striking. We conclude with some further conjectures.

Theorem 11 (Roth's Theorem).   If A is a set of positive integers with positive upper density, then A contains a three-term arithmetic progression.

Proof (Szemerédi). We call $M \subseteq N$ a k-cube if there exist $a > 0$ and $d_1, \ldots, d_k > 0$ so that

$$M = M(a: d_1, \ldots, d_k) = \left\{ a + \sum_{i=1}^{k} \varepsilon_i d_i : \varepsilon_i = 0, 1 \right\}.$$

Cube Lemma.   Let n, α, k be such that the sequence $\alpha = \alpha_0, \alpha_1, \ldots, \alpha_k$ satisfying $\alpha_{i+1} = \left\{ \binom{\alpha_i}{2} / (n-1) \right\}$ has $\alpha_k \geq 1$. If $A \subseteq [n]$ with $|A| = \alpha$, there exists a k-cube $M \subset A$. In particular, if $|A| = cn$, c fixed, there exists a k-cube $M \subset A$ with $k = \log \log n + O(1)$.

Proof. Among the $\binom{\alpha}{2}$ positive differences $a' - a$ with $a, a' \in A$, at least $\binom{\alpha}{2} / (n-1)$ must be equal. Setting d_1 equal to the most frequently occurring difference, and $A_1 = \{ a \in A : a + d_1 \in A \}$, we have

$$A_1 \subseteq A, \qquad d_1 + A_1 \subseteq A, \qquad |A_1| \geq \alpha_1.$$

Applying this argument to A_1 yields d_2, A_2 with

$$A_2 \subseteq A_1, \qquad d_2 + A_2 \subseteq A_1, \qquad |A_2| \geq \frac{\binom{|A_1|}{2}}{n-1} \geq \alpha_2$$

and, by induction, d_i, A_i with

$$A_i \subseteq A_{i-1}, \qquad d_i + A_i \subseteq A_{i-1}, \qquad |A_i| \geq \alpha_i.$$

Since $\alpha_k \geq 1$, there exists $a \in A_k$. Now $M(a: d_i, \ldots, d_k) \subseteq A_{i-1}$ by a simple backward induction on i so that

$$M = M(a: d_1, \ldots, d_k) \subseteq A$$

is the desired k-cube.

The analytic result (we really need only that k approaches infinity with n) is indicated by noting that $\alpha_{i+1} \sim \alpha_i^2/2n$ so that $\log\log(n/\alpha_i) \sim i + O(1)$. We omit the details.

Historical Note. In 1892 D. Hilbert proved that, for any $k \geq 1$, if N is finitely colored then there exists in one color infinitely many translates of a k-cube.

For every $l > 0$ let $S(l)$ denote the largest number of elements of $[1, l]$ that can be chosen so that no three-term AP is formed. Our objective, then, is to show that $\lim S(l)/l = 0$. The function S satisfies

$$S(l_1 + l_2) \leq S(l_1) + S(l_2)$$

as we may split $[1, l_1 + l_2]$ into disjoint intervals of sizes l_1 and l_2. Such functions are called *subadditive*, and we require a general lemma on them.

Subadditivity Lemma. If $S: N \to R^+$ is subadditive, then $\alpha = \lim S(n)/n$ exists and $S(n)/n \geq \alpha$ for all $n \in N$.

Proof. Set $\alpha = \lim \sup S(n)/n$. Let $n \in N$. Any $x \in N$ may be written as $x = qn + r, 0 \leq r < n$. Then

$$S(x) \leqslant S((q+1)n) \leqslant (q+1)S(n)$$

so that

$$\frac{S(x)}{x} \leqslant \frac{(q+1)S(n)}{qn}$$

Thus $\alpha \leqslant S(n)/n$. Since n was arbitrary, $\alpha \leqslant \lim\inf S(n)/n$ so that $\alpha = \lim S(n)/n$.

We prove Roth's theorem by a *reductio ad absurdum*. Assuming its negation, there exists $c > 0$ so that $c = \lim S(l)/l$ and $S(l) \geqslant cl$ for all l. Let $\varepsilon > 0$ be very small, $\varepsilon = 10^{-10}c^2$ to be specific. Let l_0 be such that

$$c \leqslant \frac{S(l)}{l} < c + \varepsilon \qquad \text{for all } l \leqslant l_0 .$$

Let l be sufficiently large so that all asymptotic approximations we shall make are justified and so that (looking ahead) $0.01c^2 \log\log l > l_0$. Let $A \subseteq [l]$, $|A| \geqslant cl$, contain no three-term AP.

Let us show the existence of a large-dimensional cube $M \subseteq A$ of small diameter not near the edges of $[l]$. On $[1, 0.49l]$ and $[0.5l, l]$ A has a total of at most $0.99l(c + \varepsilon)$ elements. Since $|A| \geqslant cl$ and ε is so small, A has density $> c/2$ on $(0.49l, 0.5l)$. (In fact, A has density "nearly" c on every "large" interval.) We split $(0.49l, 0.5l)$ into disjoint subintervals of size $l^{1/2} + O(1)$. On one of these, A has density $\geqslant c/2$. In that interval there exists a k-cube M so that

(i) $M = M(a: d_1, \ldots, d_k) \subseteq A$,
(ii) $k = \log\log l^{1/2} + O(1) = \ln\ln l + O(1)$,
(iii) $M \subset (0.49l, 0.5l)$ (a convenience),
(iv) $d_i \leqslant 2l^{1/2}, 1 \leqslant i \leqslant k$.

Set $M_{-1} = \{a\}$, $M_i = M(a: d_1, \ldots, d_{i-1})$ for $0 \leqslant i \leqslant k$. Set

$$N_i = \{2m - x : x \in A, x < a, m \in M_i\} .$$

The $y = 2m - x \in N_i$ are third terms of progressions $\{x, m, y\}$ with $x, m \in A$. Hence $A \cap N_i = \emptyset$. A has density at least $c/2$ on $[1, 0.49l]$ so that

$$|N_i| \geqslant |N_{-1}| = |A \cap (1, a)| \geqslant 0.245cl .$$

Since $M_{i+1} = M_i \cup (M_i + d_i)$, $N_{i+1} = N_i \cup (N_i + 2d_i)$. The N_i form an ascending sequence with $|N_k| \leq l$. Thus, critically, there exists i, which we fix so that

$$|N_{i+1} - N_i| < \frac{l}{k}.$$

Let us call an AP with difference $2d_i$ a block. There is a bijective correspondence between maximal blocks $\{x, x + 2d_i, \ldots, x + s(2d_i)\}$ of N_i and elements $x + (s + 1)(2d_i)$ of $N_{i+1} - N_i$. Thus N_i may be partitioned into at most l/k blocks. We split $[l]$ into the $2d_i$ residue classes modulo $2d_i$. On each class, if N_i is partitioned into t blocks then $[l] - N_i$ is partitioned into at most $t + 1$ blocks (the gaps plus the ends). In toto, $[l] - N_i$ is partitioned into at most

$$\frac{l}{k} + 2d_i = \frac{l}{\log \log l}(l + o(1))$$

(recall that $d_i < 2l^{1/2}$) blocks.

Now we may begin. We call a block of $[l] - N_i$ small if it is $< 0.01c^2 \log \log l$, and large otherwise. All of the small blocks together have at most only $0.01c^2 l + o(l)$ elements. We have defined l so that A has density $< c + \varepsilon$ on every large block, hence on their union. Every element of A (since $A \cap N_i = \varnothing$) is in either a large block or a small block:

$$\begin{aligned}
|A| &= |A \cup ([l] - N_i)| \\
&< (c + \varepsilon)(l - |N_i|) + 0.01c^2 l + o(l) \\
&< cl - c(0.245cl) + \varepsilon l + 0.01c^2 l + o(l) \\
&< cl,
\end{aligned}$$

contradicting our assumption. Hence Roth's theorem is proved.

Proof of Theorem 11 (Roth). Let $S(n)$ be as in the preceding proof. Set $c = \lim S(n)/n$. We may assume, as before, that this limit exists, $c > 0$, and $S(n)/n \geq c$ for all n. Let $\varepsilon = 10^{-10}c^2$, and let m be large enough so that

$$c \leq \frac{S(n)}{n} < c + \varepsilon \qquad \text{for } 2m + 1 \leq n.$$

Let $2N$ be sufficiently large so that the asymptotic inequalities we shall write are valid. Let $A \subseteq [2N]$, $|A| \geq c(2N)$ contain no three-term AP. It will be convenient to let $u_1, \ldots, u_r$ denote all the elements of A, and $2v, \ldots, 2v_s$, the even elements of A. Then

$$c(2N) \leq r \leq (c + \varepsilon)(2N), \qquad (c - \varepsilon)N \leq s \leq (c + \varepsilon)N,$$

the latter as (with $N \geq 2m + 1$) A can have density at most $c + \varepsilon$ on the odd or even numbers. We define two complex valued functions:

$$f(\alpha) = \sum_{i=1}^{r} e(\alpha u_i), \qquad e(x) = e^{2\pi\sqrt{-1}x} = \cos x + \sqrt{-1}\sin x,$$

$$g(\alpha) = \sum_{j=1}^{s} e(\alpha v_j).$$

Let Σ^* denote throughout the sum over $\alpha = i/2N, 0 \leq i \leq 2N - 1$. (In Roth's original paper, equivalent integrals were used.) If $u = t/2N$, where $|t| < 2N$ and t is integral, then

$$\Sigma^* e(\alpha u) = \begin{cases} 2N & \text{if } u = 0, \\ 0 & \text{if } u \neq 0. \end{cases}$$

We use this to sieve for progressions in A:

$$\Sigma^* f(\alpha)g^2(-\alpha) = \sum_{i=1}^{r} \sum_{j=1}^{s} \sum_{k=1}^{s} \Sigma^* e(\alpha(u_i - v_j - v_k))$$

$$= s(2N) < 3cN^2 \tag{1}$$

as $u_i - v_j - v_k = 0$ implies that $\{2v_j, u_i, 2v_k\}$ forms a progresion in A except in the s trivial cases $2v_j = u_i = 2v_k$. The functions f, g "spike" at $\alpha = 0$ with

$$f(0)g^2(0) = rs^2 > c^3N^3. \tag{2}$$

Our main effort will be to bound $|f(\alpha)|$ when $\alpha \neq 0$. First, we require a general theorem on Diophantine approximation: For α arbitrary, $M > 0$ integral, there exist p, q integral with

$$\alpha = \frac{p}{q} + \beta, \qquad 1 \leq q \leq M \quad \text{and} \quad q|\beta| \leq M^{-1}. \tag{3}$$

(We outline the proof. Calculated modulo 1, two of the $M + 1$ numbers $i\alpha, 0 \le i \le M$, are within M^{-1}, say $i\alpha$ and $j\alpha$. Set $q = |i - j|$ so that $q\alpha$ is within M^{-1} of an integer p. Thus $|q\alpha - p| \le M^{-1}$; now divide by q.)

Second, from elementary calculus

$$\left| \tfrac{1}{2}(e(x) + e(-x)) - 1 \right| = |\cos x - 1| \le \frac{x^2}{2},$$

from which it readily follows that

$$\left| \frac{1}{2m+1} \sum_{|i| \le m} e(i\gamma) - 1 \right| \le \frac{(m\gamma)^2}{2},$$

and thus, multiplying through by $e(\alpha)$, we obtain

$$\left| \frac{1}{2m+1} \sum_{|i| \le m} e(\alpha + i\gamma) - e(\alpha) \right| \le \frac{(m\gamma)^2}{2}.$$

Set $M = [N^{1/2}]$. (Any M in a wide range will do.) For $\alpha \ne 0$, let p, q, β satisfy (3). Then

$$e(\alpha(u + iq)) = e(\alpha u + i(\beta q))$$

so that

$$\left| e(\alpha u) - \frac{1}{2m+1} \sum_{|i| \le m} e(\alpha(u + iq)) \right| \le \frac{(m\beta q)^2}{2}$$

$$\le \frac{m^2 M^{-2}}{2}.$$

Now we may "smear" $f(\alpha)$:

$$\left| \sum_{u \in A} e(\alpha u) - \frac{1}{2m+1} \sum_{u \in A} \sum_{|i| \le m} e(\alpha(u + iq)) \right| \le \frac{|A| m^2 M^{-2}}{2}$$

$$\le \frac{m^2 N M^{-2}}{2}.$$

Let us rewrite

$$\frac{1}{2m+1} \sum_{u \in A} \sum_{|i| \le m} e(\alpha(u + iq)) = \sum_{s=0}^{2N-1} e(\alpha s) \frac{|W_s \cap A|}{2m+1},$$

where $W_s = \{s + iq : |i| \leqslant m\}$, calculated modulo $2N$. Our objective is to show that $|W_s \cap A| \sim c(2m+1)$, on the average, so that the above sum is small. Set

$$E_s = \frac{|W_s \cap A|}{2m+1} - c .$$

For $mq < s < N - mq$, W_s forms an AP of length $2m+1$ in $[2N]$. Thus $|W_s \cap A| \leqslant (2m+1)(c+\varepsilon)$ so that $E_s \leqslant \varepsilon$. For the $2mq$ other values of s we have the trivial bound $E_s \leqslant 1$. We have no good lower bound on E_s, as $W_s \cap A = \varnothing$ is quite possible. Each $a \in A$ appears in exactly $2m+1$ sets W_s so, double counting,

$$\sum_{s=0}^{2N-1} \frac{|W_s \cap A|}{2m+1} = |A| \frac{2m+1}{2m+1} = |A| .$$

Hence the average value of E_s,

$$\frac{\sum_{s=0}^{2N-1} E_s}{2N} = \frac{|A|}{2N} - c ,$$

is nonnegative. Let Σ^+ denote a summation restricted to positive terms. Then

$$\sum_{s=0}^{2N-1} |E_s| \leqslant 2 \sum_{s=0}^{2N-1} E_s \leqslant 2(2N\varepsilon + 2mq)$$

$$\leqslant 4\varepsilon N + 4mM \leqslant 5\varepsilon N$$

for N sufficiently large. For $\alpha \neq 0$, $\Sigma_{s=0}^{2N-1} e(\alpha s) = 0$ so that

$$\left| \sum_{s=0}^{2N-1} e(\alpha s) \frac{|W_s \cap A|}{2m+1} \right| = \left| \sum_{s=0}^{2N-1} e(\alpha s) E_s \right|$$

$$\leqslant \sum_{s=0}^{2N-1} |E_s| \leqslant 5N\varepsilon ,$$

and

$$|f(\alpha)| \leqslant \frac{m^2 M^{-2}}{2} N + 5N\varepsilon$$

$$\leqslant 6\varepsilon N$$

(for N sufficiently large) provides the desired upper bound.

Now we complete the proof. As $g(\alpha)$ and $g(-\alpha)$ are complex conjugates, $|g^2(-\alpha)| = g(\alpha)g(-\alpha)$ so that

$$\Sigma^* |g^2(-\alpha)| = \Sigma^* g(\alpha)g(-\alpha)$$

$$= \sum_{j=1}^{s} \sum_{k=1}^{s} \Sigma^* e(\alpha(v_j - v_k))$$

$$= 2Ns \leqslant 3cN^2$$

as the inner summation in nonzero exactly when $v_j = v_k$. We bound

$$\left| \sum_{\alpha \neq 0}^{*} f(\alpha)g^2(-\alpha) \right| \leqslant \left(\max_{\alpha \neq 0} |f(\alpha)| \right) \sum^* |g^2(-\alpha)|$$

$$\leqslant 18\varepsilon cN^3 .$$

Combining this with (1) and (2) gives

$$c^2N^3 \leqslant f(0)g^2(0)$$

$$< \left| \sum^* f(\alpha)g^2(-\alpha) \right| + \left| \sum_{\alpha \neq 0}^{*} f(\alpha)g^2(-\alpha) \right|$$

$$\leqslant 3cN^2 + 18\varepsilon cN^3 ,$$

which is impossible for N sufficiently large.

In the first edition we asked if the following result holds. It is a strengthened version of a conjective of L. Moser and would bear the same relation of the Hales–Jewett theorem that Szemerédi's theorem bears to van der Waerden's theorem.

Conjecture. For all $t \geqslant 2$ and $\varepsilon > 0$ there exists $N = N(t, \varepsilon)$ so that, if $n \geqslant N$ and $S \subseteq C_t^n$ has at least εt^n elements, there S contains a line.

This conjecture holds for $|A| = 2$, by Sperner's lemma on maximal families of incomparable subsets of a set. it would clearly imply Szemerédi's theorem. Very recently H. Furstenberg and G. Katznelson have reported proving this conjecture, using powerful extensions of the methods of Section 6.1. Their proof has not yet appeared.

In connection with Szemerédi's theorem, we remark that Erdös has conjectured the following stronger result (for proof of which he currently offers 3000 U.S. dollars).

Conjecture E. If A is a set of positive integers satisfying

$$\sum_{a \in A} \frac{1}{a} = \infty \, ,$$

then A contains arbitrarily long APs.

An affirmative answer to Conjecture E would imply the existence of arbitrarily long APs of primes.

2.6 THE SHELAH PROOF

In 1987 the Israeli logician Saharon Shelah shocked the combinatorial world by finding a fundamentally new proof of the Hales–Jewett theorem, and hence of van der Waerden's theorem. Shelah's proof gives upper bounds for the associated functions $HJ(r, t)$ and $W(k, r)$ that are fundamental improvements over the previous proofs—a topic we defer to Section 2.7. Shelah's proof, unlike that of van der Waerden, does not require a double induction. The number of colors r may be considered fixed, but arbitrary, throughout the proof. Indeed, the reader may set $r = 2$ throughout this section without any loss of the depth and ingenuity of the argument. Most surprisingly, Shelah's proof does not require an elaborate technical apparatus but rather is totally elementary in nature.

In this section we give Shelah's proof of the Hales–Jewett theorem. The proof will be totally self-contained.

It is convenient to make a slight change of notation from previous sections and let the underlying alphabet of t symbols be denoted $\{1, \ldots, t\}$. We define

$$C_t^n = \{(x_1, \ldots, x_n): x_i \in \{1, \ldots, t\}\}$$

DEFINITION. $L \subset C_t^n$ is a *Shelah line* if there is an ordering of L by $l_1, l_2, \ldots, l_t$ with $l_k = (x_{k1}, \ldots, x_{kn})$ and there exist i, j with $0 \le i < j \le n$ so that

$$x_{ks} = \begin{cases} t-1, & s \le i \, , \\ k, & i < s \le j \, , \\ t, & j < s \, . \end{cases}$$

Example. In all examples in this section we shall set $t = 26$ and associate $\{1, \ldots, 26\}$ with the English alphabet $\mathscr{A} = \{A, B, C, \ldots, X, Y, Z\}$ under the usual ordering. With $n = 9$, $i = 2$, $j = 5$ the Shelah line L has the form

$$
\begin{array}{ccccccccc}
Y & Y & Z & Z & Z & Z & Z & Z & Z \\
Y & Y & Y & Y & Y & Z & Z & Z & Z \\
Y & Y & X & X & X & Z & Z & Z & Z \\
& & & & \vdots & & & & \\
Y & Y & B & B & B & Z & Z & Z & Z \\
Y & Y & A & A & A & Z & Z & Z & Z
\end{array}
$$

Here, and throughout this section, parentheses and commas have been removed for clarity.

We call $l = (x_1, \ldots, x_n) \in C_t^n$ a *Shelah point* if it belongs to some Shelah line. A Shelah point's coordinates must consist of a (possibily empty) block of $t - 1$'s followed by a nonempty constant block followed by a (possibly empty) block of t's. Observe that a Shelah point is determined by i, j, k with $0 \le i < j \le n$ and $1 \le k \le t$ so that C_t^n contains at most $\binom{n+1}{2} t$ Shelah points.

Now suppose $n_1, \ldots, n_s$ are given, $n = n_1 + \cdots + n_s$ and associate C_t^n with $C_t^{n_1} \times C_t^{n_2} \times \cdots \times C_t^{n_s}$. For $1 \le j \le s$ let L_j be a Shelah line of $C_t^{n_j}$. Then we call $L_1 \times \cdots \times L_s$ a *Shelah s-space* of C_t^n.

Example. With $n_1 = 5$, $n_2 = 9$

$$
\{Y \ \alpha \ \alpha \ Z \ Z \ Y \ Y \ \beta \ \beta \ \beta \ Z \ Z \ Z \ Z : \alpha, \beta \in \mathscr{A}\}
$$

forms a Shelah plane.

Let $\varphi \colon L_1 \times \cdots \times L_s \to C_t^s$ denote the canonical isomorphism given by setting $\varphi(\xi) = \alpha_1 \cdots \alpha_s$ where α_j is the value of the moving coordinates in the jth block. In the example above

$$
\varphi(Y \ \alpha \ \alpha \ Z \ Z \ Y \ Y \ \beta \ \beta \ \beta \ Z \ Z \ Z \ Z) = \alpha\beta
$$

DEFINITION. A coloring χ of C_t^s is called *fliptop* if it has the following property: Let P, Q be any two points of C_t^s that have exactly the same coordinates except in one position and suppose that in that position they have values $t - 1$ and t. Then P and Q have the same color.

Example. With $s = 5$ *BAZOO* and *BAYOO* have the same color. Also *ZEZAK, ZEYAK, YEZAK*, and *YEYAK* have the same color and *YYYY, ZYYYY, ZZYYY, ZZZYY, ZZZZY* have the same color. No conditions are made on the color of *ERDOS, TETEL*, or any word with neither Y nor Z.

DEFINITION. Let $L_1 \times \cdots \times L_s$ be a Shelah s-space with $\varphi: L_1 \times \cdots \times L_s \to C_t^s$ the canonical isomorphism. A coloring χ of $L_1 \times \cdots \times L_s$ is called fliptop if the derived coloring χ' of C_t^s given by $\chi'(P) = \chi[\varphi^{-1}(P)]$ is fliptop.

Example. With the Shelah plane given above:

$$
\begin{array}{cccccccccccccc}
Y & Y & Y & Z & Z & Y & Y & Y & Y & Y & Z & Z & Z & Z \\
Y & Z & Z & Z & Z & Y & Y & Y & Y & Y & Z & Z & Z & Z \\
Y & Y & Y & Z & Z & Y & Y & Z & Z & Z & Z & Z & Z & Z \\
Y & Z & Z & Z & Z & Y & Y & Z & Z & Z & Z & Z & Z & Z
\end{array}
$$

will have the same color.

The condition for a Shelah line L to be fliptop under χ is particularly simple: We require only that the final and penultimate points of L have the same color.

Example. The Shelah line given above is fliptop if

$$
Y \quad Y \quad Y \quad Y \quad Y \quad Z \quad Z \quad Z \quad Z
$$

and

$$
Y \quad Y \quad Z \quad Z \quad Z \quad Z \quad Z \quad Z \quad Z
$$

have the same color.

Lemma. Assume $n \geq c$. let C_t^n be c-colored arbitrarily. Then there exists a fliptop Shelah line.

Proof. For $0 \leq i \leq n$ define $P_i = (x_{i1}, \ldots, x_{in})$ by

$$
x_{ij} = \begin{cases} t - 1, & j \leq i, \\ t, & j > i. \end{cases}
$$

As $n + 1 > c$ by the Pigeon-Hole principle some two of these points P_i, P_j have the same color. These are the last two points of the Shelah line $l_1, \ldots, l_t$ with $l_t = (x_{k1}, \ldots, x_{kn})$ defined by

$$
x_{ks} = \begin{cases} t - 1, & s \leq i \\ k, & i < s \leq j, \\ t, & j < s. \end{cases}
$$

Example. With $n = c = 5$ some two of the points $ZZZZZ$, $YZZZZ$, $YYZZZ$, $YYYZZ$, $YYYYZ$, and $YYYYY$ have the same color. If, say, $YZZZZ$ and $YYYZZ$ are the same color then $L = \{Y\alpha\alpha ZZ : \alpha \in \mathcal{A}\}$ is fliptop.

Mighty oaks form little acorns grow, though it did take 60 years and Saharon Shelah to find the right acorn!

Theorem 12. Let r, s, t be fixed positive integers. Define $n_1, \ldots, n_s$ by

$$n_1 = r^{t^{s-1}}$$

$$n_2 = r^{\binom{n_1+1}{2} t^{s-1}}$$

and, in general, with n_i having been defined set

$$A_i = \left[\prod_{j \le i} \binom{n_j + 1}{2} \right] t^{s-1}$$

and

$$n_{i+1} = r^{A_i}, \qquad 1 \le i < s.$$

Set $n = n_1 + \cdots + n_s$. let an arbitrary r-coloring χ of C_t^n be given. Then there is a fliptop Shelah s-space.

Proof. We associate C_t^n with $C_t^{n_1} \times \cdots \times C_t^{n_s}$ and write a point $y \in C_t^n$ as $y = y_1, \ldots, y_s$ where $y_j \in C_t^{n_j}$. We define an equivalence relation $\equiv$ on $C_t^{n_s}$ by setting

$$y_s \equiv y_s' \text{ iff } \chi(y_1, \ldots, y_{s-1}, y_s) = \chi(y_1, \ldots, y_{s-1}, y_s')$$

$$\text{for } all \text{ Shelah points } y_1, \ldots, y_{s-1}$$

There are at most A_{s-1} choices for $y_1 y_2, \ldots, y_{s-1}$ and hence at most $n_s = r^{A_{s-1}}$ equivalence classes. The equivalence relation $\equiv$ may be considered an n_s-coloring $\hat{\chi}$ of $C_t^{n_s}$. Applying the lemma there exists a Shelah line $L_s \subset C_t^{n_s}$, fliptop under $\hat{\chi}$.

Suppose, by reverse induction, $L_s, L_{s-1}, \ldots, L_{i+1}$ have been found. We define an equivalence relation $\equiv$ on $C_t^{n_i}$ by setting $y_i \equiv y_i'$ if and only if

$$\chi(y_1, \ldots, y_{i-1}, y_i, z_{i+1}, \ldots, z_s) = \chi(y_1, \ldots, y_{i-1}, y_i', z_{i+1}, \ldots, z_s)$$

for all Shelah points $y_1, \ldots, y_{i-1}$ and all choices of $z_{i+1} \in L_{i+1}, \ldots, z_s \in L_s$. There are $\binom{n_j + 1}{2} t$ choices for each y_j, $1 \leq j \leq i-1$. There are only t choices for each z_j, $i + 1 \leq j \leq s$ as the lines $L_{i+1}, \ldots, L_s$ have already been determined. (This is an absolutely critical juncture in the proof as we cannot have n_i depend on the *later* values $n_{i+1}, \ldots, n_s$.) Altogether there are A_{i-1} choices of $y_1, \ldots, y_{i-1}, z_{i+1}, \ldots, z_s$. Hence there are at most $n_i = r^{A_{i-1}}$ equivalence classes so we may consider $\equiv$ as an n_i-coloring $\hat{\chi}$ of $C_t^{n_i}$. Applying the lemma there exists a Shelah line $L_i \subset C_t^{n_i}$, fliptop under this $\hat{\chi}$.

We claim that $L_1 \times \cdots \times L_s$ is the desired fliptop Shelah s-space. Fix $i, 1 \leq i \leq s$ and let y_i, y_i' be the last two points of L_i. By construction $y_i \equiv y_i'$ and so

$$\chi(y_1, \ldots, y_{i-1}, y_i, z_{i+1}, \ldots, z_s) = \chi(y_1, \ldots, y_{i-1}, y_i', z_{i+1}, \ldots, z_s)$$

for all Shelah points $y_1, \ldots, y_{i-1}$ and all $z_{i+1} \in L_{i+1}, \ldots, z_s \in L_s$. But for $1 \leq j < i$ all $y_j \in L_j$ are surely Shelah points and so

$$\chi(z_1, \ldots, z_{i-1}, y_i, z_{i+1}, \ldots, z_s) = \chi(y_1, \ldots, z_{i-1}, y_i', z_{i+1}, \ldots, z_s)$$

for all $z_j \in L_j$, $1 \leq j \leq s$, $j \neq i$, completing the proof.

Example. $r = 2$, $s = 2$, $t = 26$. Set $n_1 = 2^{26}$, $A_1 = \binom{2^{26} + 1}{2} 2^{26}$, $n_2 = 2^{A_1}$, $n = n_1 + n_2$. Each point of C_{26}^n may be uniquely written in the form xy with $x \in C_{26}^{n_1}$, $y \in C_{26}^{n_2}$.

$$\overset{\text{\tiny a}}{\underset{x}{\underline{\qquad\qquad}}} \quad \underset{y}{\underline{\qquad\qquad\qquad\qquad}}$$

We first find $y', y'' \in C_{26}^{n_2}$, each of the form $Y, \ldots, Y, Z, \ldots, Z$ so that xy' and xy'' have the same color for all Shelah points $x \in C_{26}^{n_1}$. These points y', y'' lie on a Shelah line L_2

$$
\begin{array}{lll}
Y\text{------}Y\ Z\text{------}Z\ Z\text{------------}Z = y'' \\
Y\text{------}Y\ Y\text{------}Y\ Z\text{------------}Z = y' \\
Y\text{------}Y\ X\text{------}X\ Z\text{------------}Z \\
\qquad\qquad\qquad \vdots \\
Y\text{------}Y\ A\text{------}A\ Z\text{------------}Z
\end{array}
$$

$$L_2$$

Now we find $x', x'' \in C_{26}^{n_1}$, each of the form $Y, \ldots, YZ, \ldots, Z$ so that $x'y$ and $x''y$ have the same color for all $y \in L_2$. These points x', x'' lie on a Shelah line L_1.

$$
\begin{array}{l}
x'' = Y\text{———}Y\ Z\text{———}Z\ Z\text{———}Z \\
x' = Y\text{———}Y\ Y\text{———}Y\ Z\text{———}Z \\
Y\text{———}Y\ X\text{———}X\ Z\text{———}Z \\
\vdots \\
Y\text{———}Y\ A\text{———}A\ Z\text{———}Z
\end{array}
\qquad
\begin{array}{l}
Y\text{———}Y\ Z\text{———}Z\ Z\text{———}Z = y'' \\
Y\text{———}Y\ Y\text{———}Y\ Z\text{———}Z = y' \\
Y\text{———}Y\ X\text{———}X\ Z\text{———}Z \\
\vdots \\
Y\text{———}Y\ A\text{———}A\ Z\text{———}Z
\end{array}
$$

$$\underbrace{}_{L_1} \qquad\qquad \underbrace{}_{L_2}$$

For any $y \in L_2$, $x'y$ and $x''y$ have the same color. For any $x \in L_1$ (indeed, for any Shelah point $x \in C_{26}^{n_1}$) xy' and xy'' have the same color. Hence $L_1 \times L_2$ is a Shelah plane.

Lemma. Let $s = HJ(r, t-1)$ be such that given any r-coloring of C_{t-1}^s there exists a monochromatic line. Then under any fliptop r-coloring of C_t^s there exists a monochromatic line.

Proof. Restricting the domain to $C_{t-1}^s \subset C_t^s$ there is a monochromatic line $l_1, \ldots, l_{t-1}$. Let l_t be the point of C_t^s given by setting all the moving coordinates of the line equal to t. Then $l_1, \ldots, l_{t-1}, l_t$ is a line in C_t^s. The point l_t may be derived from l_{t-1} by changing a subset of the coordinate values (namely, on the moving coordinates) from $t-1$ to t. As the coloring is fliptop each such change on a single coordinate preserves the color and hence any sequence of such changes preserves the color so that l_{t-1} *and* l_t *have the same color.* As $l_1, \ldots, l_{t-1}$ already have the same color the set $l_1, \ldots, l_{t-1}, l_t$ forms a monochromatic line in C_t^s.

Example. Suppose that with $t = 26$, $s = 3$ under a fliptop coloring ABA, BBB, $CBC, \ldots, XBX$, YBY had the same color. The YBY, ZBY, ZBZ would have the same color so $\{\alpha B\alpha : \alpha \in \mathcal{A}\}$ would form a monochromatic line.

Theorem 13 (The Hales–Jewett Theorem). For all r, t there exists $n = HJ(r, t)$ so that if C_t^n is r-colored there exists a monochromatic line.

Proof (Shelah). We fix r and use induction on t. Trivially we may take $HJ(r, 1) = 1$. Suppose, by induction, $s = HJ(r, t-1)$ exists. Let n be given by Theorem 12. Given an r-coloring χ of C_t^n there is a fliptop Shelah s-space $L_1 \times \cdots \times L_s$. Define the derived coloring χ' of C_t^s by $\chi'(y) = \chi(\varphi^{-1}(y))$ where $\varphi : L_1 \times \cdots \times L_s \to C_t^s$ is the canonical isomorphism. Then χ' is fliptop so by the lemma above there is a mono-chromatic line $L \subset C_t^s$. Then $\varphi^{-1}(L) \subset L_1 \times \cdots \times L_s$ is the derived monochromatic line in C_t^n.

2.7 EEEEENORMOUS UPPER BOUNDS

Why is Shelah's proof of the Hales–Jewett theorem considered an improvement of fundamental importance. The answer comes from examining the growth rates of the functions $HJ(r, t)$ given by the proofs of van der Waerden and Shelah. For convenience we shall look particularly at the case $r = 2$. The functions involved grow so rapidly that we must first discuss a special language—called the Ackermann hierarchy—devised by logicians to deal with rapidly growing functions.

The Ackermann hierarchy is a sequence of functions $f_1, f_2, \ldots$, with domain and range the positive integers. (There are several equivalent formualtions in the literature; we have chosen a formulation hopefully more readily comprehensible to mathematicians.) The first function, f_1, we call DOUBLE and is defined simply by

$$f_1(x) = \text{DOUBLE}(x) = 2x .$$

The second function, f_2, we call EXPONENT and may be defined by

$$f_2(x) = \text{EXPONENT}(x) = 2^x .$$

More critical, however, is that we may derive EXPONENT from DOU-BLE as follows: To find EXPONENT (x) start at 1 and apply DOUBLE x times. It is this notion of iteration that allows us to describe very rapidly growing functions. The third function, f_3, we call TOWER and is derived from EXPONENT in the same way: To find TOWER(x) start at 1 and apply EXPONENT x times. TOWER(x) may be written $2^{2^{\cdot^{\cdot^{\cdot^2}}}}$ with x twos in the "tower," hence the name. More generally, and formally, we define f_{i+1} by

$$f_{i+1}(x) = f_i^{(x)}(1)$$

where $f^{(x)}$ denotes the xth iterate of f. Alternatively, we define f_{i+1} inductively by

$$f_{i+1}(1) = 2 ,$$
$$f_{i+1}(x + 1) = f_i[f_{i+1}(x)] .$$

Note that this is really a double induction. By induction on i, we define the function f_i. With the function f_i already defined we then define $f_{i+1}(x)$ by induction on x.

The first few values of $f_i(x)$ are given in Table 2.1. Notice that $f_3(5) = 2^{65536}$ is already a number with nearly 20,000 decimal digits. In comparison, a googol has only 100 digits. The number $f_3(6)$ then has $(\log_{10} 2)f_3(5)$ decimal digits, far larger than a googolplex, well beyond any conceivable physical interpretation. We call f_4 the WOW function. This fanciful description comes from trying to grasp the magnitude of $f_4(4)$—a tower of twos of size 65,536—what can we say but "oh wow!"

Diagonalization allows an even faster growing function. The Ackermann function, denoted by f_ω or ACKERMANN, is defined by

$$f_\omega(x) = \text{ACKERMANN}(x) = f_x(x) .$$

A simple induction shows that $f_i(x)$ is monotone in both x and i. For any n if $x \geqslant n$ then $\text{ACKERMANN}(x) = f_x(x) \geqslant f_n(x)$. That is, ACKERMANN grows more rapidly than any of the f_n. Logicians can prove that ACKERMANN grows more rapidly than any primitively recursive function—which means, roughly, that a double induction is essential for its definition. (To go beyond ACKERMANN, see Section 6.3.)

We say that a function $g(x)$ is a level i function (including $i = \omega$) if there are $c', c'' > 0$ so that for x sufficiently large

$$f_i(c'x) < g(x) < f_i(c''x) .$$

For $i = 1, 2, 3, 4, \omega$ we use the words *linear, exponential, towerian, wowzer,* and *ackermanic* (the last coined by John Conway) to describe $g(x)$.

Table 2.1 Beginnings of the Ackermann Hierarchy

		1	2	3	4	5	
DOUBLE	f_1	2	4	6	8	10	12 $\cdots$
EXPONENT	f_2	2	4	8	16	32	64 $\cdots$
TOWER	f_3	2	4	16	65536	2^{65536}	$\vdots$
WOW	f_4	2	4	65536	WOW!	$\vdots$	
	f_5	2	4	WOW!	$\vdots$		
ACKERMANN	f_ω	2	4	16	WOW!		

We may now describe succinctly the breakthrough given by the Shelah proof. *All previous proofs to the van der Waerden or Hales–Jewett theorems gave as an upper bound* (with $r = 2$ colors), *an ackermanic function; Shelah's proof gives a wowzer function.*

Let's examine the arguments of Sections 2.1 and 2.6 in some detail. Let $W_k(r)$ be the value n given by the proof of Section 2.1 so that if $\{1, \ldots, n\}$ is r-colored there exists a monochromatic k-term AP. We took $W_2(r) = r + 1$. To find $W_{k+1}(r)$ we first set

$$c_1 = 2W_k(r) - 1$$

so that any block of length c_1 contains a monochromatic k-term progression plus a $(k + 1)$-st term. There are r^{c_1} ways to color such a block. We set

$$c_2 = 2W_k(r^{c_1}) - 1$$

so that with c_2 such blocks there would be a k-term progression of identically colored blocks plus a $(k + 1)$-st block. That is, we have a level 2 block consisting of c_2 level 1 blocks. More generally, we set

$$c_{i+1} = 2W_k(r^{c_i}) - 1, \qquad 1 \le i < r$$

so that an $(i + 1)$-level block consists of c_{i+1} i-level blocks. We stopped with a level r block and set

$$W_{k+1}(r) = c_1 c_2, \ldots, c_r$$

the number of elements in such a block. (For example, one may check that $W_3(2) = 5 \times 65 = 325$.)

We shall use some rather crude estimates to bound $W_k(2)$. First note that $f_k(1) = 2$, $f_k(2) = 4$ for all k; $f_k(r)$ is monotone in both k and r; and for $x \ge 4$, $k \ge 3$, $f_k(x) \ge f_3(x) \ge 5x^x$ with "room to spare," a euphemism for "left to the reader."

Lemma. For $r \ge 2$

$$f_3(r + 1) \ge W_3(r) \le f_3(3r - 1).$$

Proof. As $W_2(r) = r + 1$ we set

$$c_1 = 2r + 1$$

$$c_{i+1} = g(c_i) \qquad \text{with } g(x) = 2r^x + 1,$$

and $W_3(r) = c_1 \cdots c_r$. $c_1 \geq 4 = f_3(2)$. As $g(x) \geq 2^x$ by induction on i, $c_i \geq f_3(i+1)$ so that $W_3(r) \geq c_r \geq f_3(r+1)$.

The upper bound holds for $r = 2$ by inspection. For $r \geq 3$, $c_1 \leq 2^r = f_2(r) \leq f_3(r)$. As all $c_i \geq r$ when $x = c_i$

$$g(x) \leq 2x^x + 1 \leq 2^{2^x}$$

with "room to spare." Thus $c_i \leq f_3(a)$ implies $c_{i+1} \leq f_3(a+2)$. By induction $c_r \leq f_3(3r-2)$. Then

$$W_3(r) \leq c_r^r \leq 2^{c_r} \quad \text{(with room to spare)}$$
$$\leq f_3(3r-1).$$

Lemma. For $k \geq 3$, $r \geq 2$

$$f_k(r+1) \leq W_k(r) \leq f_k(5r).$$

Proof. We have shown this for $k = 3$. Now assume, by induction, that the result holds for k. Then

$$c_1 \geq W_k(r) \geq f_k(r+1)$$

by induction. For each i

$$c_{i+1} \geq W_k(r^{c_i}) \geq W_k(c_i) \geq f_k(c_i)$$

so by induction on i

$$c_i \geq f_k^{(i)}(r+1).$$

Thus

$$W_k(r) \geq c_r \geq f_k^{(r)}(r+1) \geq f_k^{(r)}(2) = f_k^{(r+1)}(1) = f_{k+1}(r+1).$$

For the upper bound

$$c_1 < 2W_k(r) < f_k[f_k(5r)]$$

as $f_k(x) > 2x$. As $c_i > r$

$$c_{i+1} \leqslant 2W_k(r^{c_i}) \leqslant 2W_k(c_i^{c_i})$$
$$\leqslant 2f_k(5c_i^{c_i}) \quad \text{(induction on } k\text{)}$$
$$\leqslant f_k\{f_k[f_k(c_i)]\}$$

as $f_k(x) \geqslant 2x$, $5x^x$. Thus $c_r \leqslant f_k^{(3r-1)}(5r)$ and

$$W_k(r) \leqslant c_r^r \leqslant f_k(c_r) \quad \text{(with room to spare)}$$
$$\leqslant f_k^{(3r)}(5r) .$$

For all $r \geqslant 2$, $f_{k+1}(r+1) \geqslant 4r$ with room to spare so that

$$W_k(r) \leqslant f_k^{(3r)}(f_{k+1}(r+1)) = f_{k+1}(4r+1) < f_{k+1}(5r) .$$

Claim. For $k \geqslant 10$,

$$\text{ACKERMANN}(k-2) \leqslant W_k(2) \leqslant \text{ACKERMANN}(k) .$$

Proof

$$W_k(2) \leqslant f_k(10) \leqslant f_k(k) = \text{ACKERMANN}(k)$$
$$W_k(2) \geqslant f_k(3) = f_{k-1}\{f_{k-1}[f_{k-1}(1)]\} = f_{k-1}(4)$$
$$= f_{k-2}(f_{k-2}\{f_{k-2}[f_{k-2}(1)]\})$$
$$= f_{k-2}(f_{k-2}(4))$$

But $f_t(4) \geqslant t$ with room to spare so

$$W_k(2) \geqslant f_{k-2}(k-2) = \text{ACKERMANN}(k-2)$$

Note that the robustness of ACKERMANN is such that even with these extremely rough bounds the value of $W_k(2)$ is found "within 2." The appearance of precision is deceptive because of the growth rate of ACKERMANN—in another sense the bounds on $W_k(2)$ are two full levels apart.

Now we turn to Section 2.6 and let $S(t)$ be that number n given by Shelah's proof so that if C_t^n is 2-colored there exists a monochromatic line. The recursion is as follows. Set $S(1) = 1$. Suppose $S(t-1) = s$ has been defined. Set

$$n_1 = 2^{t^{s-1}}$$

and for $1 \leq i < s$ set

$$n_{i+1} = 2^{A_i}$$

where

$$A_i = \left[\prod_{j \leq i} \binom{n_j + 1}{2} \right] t^{s-1} .$$

Finally, set

$$S(t) = n = n_1 + \cdots + n_s .$$

Roughly, n_i will be a tower of size i so that $S(t)$ will be a tower of size $s = S(t-1)$, hence S will be a wowzer function. When $t = 2$, $s = S(1) = 1$, $n = n_1 = 2^{s^{1-1}} = 2$. (Some two of the points 11, 12, 22 are the same color giving the monochromatic line.) When $t = 3$, $s = S(2) = 2$, $n_1 = 2^{3^{2-1}} = 8$, $A_1 = \binom{8+1}{2} 3^2 = 288$, $n_2 = 2^{288}$, $S(3) = 8 + 2^{288}$.

Claim. For $t \geq 3$

$$\mathrm{WOW}(t) \leq S(t) \leq \mathrm{WOW}(t+1) .$$

We first show the lower bound by induction. For $t = 3$ it holds by inspection. Assume it true for $t - 1$ and let $n_1, \ldots, n_s$ be as defined. $n_1 \geq 2 = \mathrm{TOWER}(1)$. As $A_i \geq n_i$, $n_{i+1} \geq 2^{n_i}$ so by induction on i, $n_i \geq \mathrm{TOWER}(i)$. Thus

$$S(t) \geq n_s \geq \mathrm{TOWER}(s) \geq \mathrm{TOWER}[\mathrm{WOW}(t-1)] = \mathrm{WOW}(t) .$$

For the upper bound we prove the stronger hypothesis $S(t) \leq \mathrm{WOW}(t+1)/6$ by induction on t. For $t = 3$ it holds by inspection—indeed $\mathrm{WOW}(4)$ was our "wow!" number. Assume the hypothesis for $t - 1$.

$$n_1 = 2^{t^{s-1}} < s^{s^s} < \mathrm{TOWER}(s)$$

with room to spare. For any i, bounding t and all n_j by n_i

$$A_i \leq n_i^s n_i^{2s} \leq n_i^{3n_i} < 2^{2^{n_i}}$$

with room to spare. If $n_i \leq \text{TOWER}(a)$ then $n_{i+1} \leq \text{TOWER}(a + 3)$. hence $n_s \leq \text{TOWER}(s + 3(s - 1)) = \text{TOWER}(4s - 2)$ and

$$S(t) = n_1 + \cdots + n_s < sn_s < 2^{n_s} \leq \text{TOWER}(4s - 1).$$

[This may be tightened to $S(t) < \text{TOWER}\ (s(1 + o(1))$, but such differences at the TOWER level evaporate at the WOW level.] By induction $s \leq \text{WOW}(t + 1)/6$ so $4s - 2 < \text{WOW}(t + 1) - 1$ and

$$S(t) \leq \text{TOWER}(\text{WOW}(t + 1) - 1) = \log_2(\text{WOW}(t + 2))$$

$$< \text{WOW}(t + 2)/6$$

completing the argument.

Striking in these detailed arguments is the robustness of a wowzer or ackermanic function. One can do "just about anything" to such a function and it retains its level. Dealing with bounds on combinatorial functions at this level requires a particular feeling for these functions— often seemingly gross improvements give no change in the function level. This robustness had led many mathematicians to speculate that the Hales–Jewett function was intrinsically ackermanic. In the first edition of this work we wrote: "Perhaps van der Waerden's function or, more naturally, the Hales–Jewett function $HJ(r, t)$ can be proven to grow very quickly. Such a proof may come from mathematical logicians; indeed, several logicians believe that a model-theoretic argument is possible. Perhaps, in some precise way, the Hales–Jewett theorem for $r = 2$ colors cannot be proved without a proof for all r." At the time we hardly imagined that a proof would come from a logician—but a proof that the Hales–Jewett function was *not* ackermanic using totally combinatorial methods!

Let $W(k)$ denote the true value of van der Waerden's function for two colors—the least n so that if $\{1, \ldots, n\}$ is 2-colored then there exists a monochromatic arithmetic progression with k-terms. The lower bounds on $W(k)$ (see Section 4.3) are exponential. Shelah's upper bound (using the translation from the Hales–Jewett theorem to van der Waerden's theorem given in Section 2.2) is wowzer. A large gap still remains. For a number of years the senior author has had a standing offer of \$1000 for a proof (or disproof) of the following. (This was finally paid to Tim Gowers.)

Conjecture. $W(k) \leq \text{TOWER}(k)$ for all k.

It was felt that Shelah's bound was such an improvement over what was previously available that he was awarded half of the offered prize. The conjecture is still open, however, and the original prize offer still stands to anyone (including Shelah) who settles it.

Further Improvement? Can Shelah's proof be "tightened" to give, say a towerian upper bound to $HJ(2, t)$. If we knew for sure we would rush to publication ourselves. Still—let's speculate!

Let $F(d, r)$ be the least n with the following property. Let χ be an arbitrary r-coloring of all d-tuples:

$$(\{x_1, y_1\}, \ldots, \{x_{i-1}, y_{i-1}\}, z_i, \{x_{i+1}, y_{i+1}\}, \ldots, \{x_d, y_d\}). \quad (*)$$

Here all values lie in $\{1, \ldots, n\}$, $1 \le i \le d$, and $x_j \ne y_j$ for all $j \ne i$. Then there exist $\{x_j, y_j\}$, $1 \le j \le d$, with $x_j \ne y_j$, all j, so that for each i

$$(\{x_1, y_1\}, \ldots, \{x_{i-1}, y_{i-1}\}, x_i, \{x_{i+1}, y_{i+1}\}, \ldots, \{x_d, y_d\})$$

and

$$(\{x_1, y_1\}, \ldots, \{x_{i-1}, y_{i-1}\}, y_i, \{x_{i+1}, y_{i+1}\}, \ldots, \{x_d, y_d\})$$

and the same color.

Claim. Let $s = HJ(2, t-1)$. Then $HJ(2, t) \le sF(s, 2^{t^{s-1}})$.

Proof. Set $n_1 = \cdots = n_s = F(s, 2^{t^{s-1}})$, $n = n_1 + \cdots + n_s$ and fix a 2-coloring of $C_t^n = C_t^{n_1} \times \cdots \times C_t^{n_s}$. To the d-tuple $(*)$ we associate a Shelah $(s-1)$-space $L_1 \times \cdots \times L_{i-1} \times w \times L_{i+1} \times \cdots \times L_s$ as follows. Let w be the string consisting of z_i $(t-1)$'s followed by $(n_i - z_i)$ t's. For $j \ne i$ let L_j be the Shelah line consisting of all strings with (letting $x_j < y_j$) $x_j(t-1)$'s followed by $y_i - x_i$ k's and then $n_i - y_i$ t's, where k runs from 1 to n. Induce a coloring of the d-tuples by the way the associated Shelah space is colored. Since the Shelah space has t^{s-1} points there are at most $2^{t^{s-1}}$ induced "colors." A family $\{x_j, y_j\}$, $1 \le j \le d$ with the given property then corresponds to a fliptop Shelah s-space $L_1 \times \cdots \times L_s$.

Any roughly exponential bound on F, for example, $F(d, r) < 2^{c \max[d, r]}$, would translate into a towerian bound on $HJ(2, t)$. The known upper bounds on $F(d, 2)$ (following the proof of Theorem 12—indeed, this was the original argument of Shelah) are towerian in d.

The function $F(2, r)$ has an interesting interpretation. Let S_n be the set of lattice points (i, j) in the plane with $1 \le i, j \le n$. Define the mesh clique graph G_n on S_n by letting two points be adjacent if they have either the same first or the same second coordinate. $F(2, r)$ is then the least n so that given any r-coloring of G_n there is a "rectangle" (i, j), (i, j'), (i', j), (i', j') so that the vertical edges [from (i, j) to (i, j') and from (i', j) to (i', j')] are the same color and the horizontal edges [from (i, j) to (i', j)

and from (i, j') to $(i', j')]$ are the same color. [The connection is given by associating $(\{i, i'\}, j$ with $\{(i, j), (i', j)\}$ and $(i, \{j, j'\})$ with $\{(i, j), (i, j')\}.]$ A polynomial upper bound to $F(2, r)$ might well lead to a towerian upper bound to $HJ(2, t)$. Even if not, it is certainly an interesting problem for its own sake.

REMARKS AND REFERENCES

§1.   Proofs of van der Waerden's theorem are found in van der Waerden [1927] (the original paper), van der Waerden [1971] (the expository account), and Graham and Rothschild [1974] (the short proof).

§2.   Hales and Jewett [1963] provide the basic reference.

§3.   Gallai's theorem is found in Rado [1933a].

§4.   Graham, Leeb, and Rothschild [1972] and Spencer [1979] provide the original and simplified proofs of the Affine and Vector Space Ramsey theorems. Cates and Hindman [1975] show that extension of the vector Space Ramsey theorem to the "infinite case" is usually false. In particular, they show that it is possible to finitely color the t-spaces of an infinite-dimensional space so that there is no infinite-dimensional monochromatic subspace.

§5.   Roth's proof appears in Roth [1952] and Roth [1953]. The full proof of Szemerédi's theorem is given in Szemerédi [1975]. Moser's conjecture appears in Moser [1970].

§6.   Shelah's proof appears in Shelah [1988].

3

Equations

3.1 SCHUR'S THEOREM

In this chapter we prove theorems of the following form: Given any finite coloring of N, there exist $x_1, \ldots, x_n \in N$ having the same color that satisfy some prescribed condition. Our prototype result was proved by I. Schur in 1916. It may perhaps be considered the earliest result in Ramsey theory.

Theorem 1 (Schur's Theorem). If N finitely colored there exist x, y, z having the same color such that

$$x + y = z .\tag{1}$$

Proof. Assume that r colors are used. Let n be such that

$$n + 1 \rightarrow (3)_r .$$

An r-coloring χ of $[n]$ induces an r-coloring χ^* of K_{n+1} on vertex set $\{0, 1, \ldots, n\}$ by $\chi^*(i, j) = \chi(|i - j|)$. There must exist a monochromatic triangle in K_n; that is, $i > j > k$ such that $\chi^*(i, j) = \chi^*(i, k)$. Setting $x = i - j$, $y = j - k$, $z = i - k$ gives $\chi(x) = \chi(y) = \chi(z)$ and $x + y = z$.

Historical Note: Schur's Paper. Schur's original paper was motivated by Fermat's Last Theorem. He actually proved the following result.

THEOREM. For all m, if p is prime and sufficiently large the equation

$$x^m + y^m = z^m$$

has a nonzero solution in the integers modulo p.

Proof. Let p be prime and sufficiently large (using the finite form of Schur's theorem) so that if $(1, \ldots, p - 1\}$ is m-colored there exist a, b, c

colored identically with $a + b = c$. Let $H = \{x^m : x \in Z_p^*\}$; H is a subgroup of Z_p^* of index $n = \gcd(m, p - 1) \le m$. The cosets of Z_p^* define an n-coloring χ of Z_p^* with the property that $\chi(a) = \chi(b)$ iff $ab^{-1} \in H$. There exist $a, b, c \in \{1, \ldots, p - 1\}$ with $\chi(a) = \chi(b) = \chi(c)$ and $a + b = c$. In Z_p

$$1 + a^{-1}b = a^{-1}c,$$

and $1, a^{-1}b$, and $a^{-1}c$ are all nonzero mth powers in Z_p.

Schur never again touched on this problem.

Although the proof of Schur's theorem is appealing in its simplicity, it will not serve for the extensions of the result. For this we need a result that strengthens both Schur's theorem and van der Waerden's theorem.

Theorem 2. For all $k, r, s \ge 1$ there exists $n = n(k, r, s)$ so that, if $[n]$ is r-colored, there exist $a, d > 0$ so that

$$\{a, a + d, a + 2d, \ldots, a + kd\} \cup \{sd\} \tag{2}$$

is monochromatic.

Proof. We use induction on r. We may clearly take $n(k, 1, s) = \max[k + 1, s]$. Let $W(t, r)$ be the minimal W such that if $[W]$ is r-colored there exists a monochromatic arithmetic progression of length t. (Here, of course, we are using van der Waerden's theorem.)

For given k, r, s we claim that we may take $n = sW(kn(k, r - 1, s), r)$. We fix an r-coloring of $[n]$. Among the first $W(kn(k, r - 1, s), r)$ integers we find a monochromatic, say red, set

$$\{a + id' : 0 \le i \le kn(k, r - 1, s)\} .$$

If, for some $j, 1 \le j \le n(k, r - 1, s)$, $sd'j$ is red then (2) is red with $d = jd'$. Otherwise $\{sd'j : 1 \le j \le n(k, r - 1, s)\}$ is $(r - 1)$-colored. Using the equivalence between colorings of $[n]$ and $sd'[n]$, we find that a monochromatic set of type (2) exists.

Corollary 3. For all $k, r, s \ge 1$ there exists $n = n(k, r, s)$ so that, if $[n]$ is r-colored, there exist $a, d > 0$ so that

$$\{a + \lambda d : |\lambda| \le k\} \cup \{sd\} \tag{3}$$

is monochromatic.

Proof. Apply Theorem 2 with $k' = 2k$, finding a', d' so that $a' + \lambda d', 0 \le \lambda \le 2k$, and sd' are the same color. Set $d = d'$ and $a = a' + kd'$.

3.2 REGULAR HOMOGENEOUS EQUATIONS (RADO'S THEOREM—ABRIDGED)

Let $S = S(x_1, \ldots, x_n)$ denote a system of equations in the variables $x_1, \ldots, x_n$. Let A be a set on which S is defined. We say that S is *r-regular* on A if, given any r-coloring of A, there exist $x_1, \ldots, x_n \in A$ (not necessarily distinct) so that $S(x_1, \ldots, x_n)$ holds and $x_1, \ldots, x_n$ are the same color. We say that S is *regular* on A if it is r-regular for all positive integers r. Schur's theorem states that the condition

$$x_1 + x_2 - x_3 = 0$$

is regular on N. Theorem 2 states that, for all k, the condition

$$x_1 = x_0 + d$$
$$x_2 = x_1 + d$$
$$\vdots$$
$$x_k = x_{k-1} + d$$

on the variables $\{x_0, \ldots, x_k, d\}$ is regular on N.

A comprehensive study of regular systems on N was the dissertation topic of one of Schur's most illustrious students—Richard Rado.

Theorem 4 (Rado's Theorem—Abridged). Let $S(x_1, \ldots, x_n)$ be given by a single linear homogeneous constraint:

$$c_1 x_1 + \cdots + c_n x_n = 0, \qquad c_i \in Z. \tag{4}$$

Then S is regular on N iff some nonempty subset of the c_i sums to zero.

Proof. We first assume, reordering for convenience, that

$$c_1 + \cdots + c_k = 0,$$

and fix a finite coloring of N. We need to find a monochromatic solution to (4). If $k = n$ we may take $x_1 = \cdots = x_n = 1$. We assume that $k < n$ and set

$$A = gcd(c_1, \ldots, c_k),$$
$$B = c_{k+1} + \cdots + c_n,$$
$$s = \frac{A}{gcd(A, B)}.$$

(If $B = 0$, $c_1 + \cdots + c_n = 0$ so that $x_1 = \cdots = x_n = 1$ gives a mono-chromatic solution.) By elementary number theory we find $t \in Z$ so that

$$At + Bs = 0$$

and $\lambda_1, \ldots, \lambda_k \in Z$ so that

$$c_1 \lambda_1 + \cdots + c_k \lambda_k = At.$$

Now (4) has a parametric solution:

$$x_i = \begin{cases} a + \lambda_i d, & 1 \leq i \leq k, \\ sd, & k < i. \end{cases} \tag{5}$$

By Corollary 3 we can find a, d so that $\{x_i : 1 \leq i \leq n\}$ is monochromatic, completing the "if" section of Theorem 4.

We illustrate this method with the equation

$$x_1 + 3x_2 - 4x_3 + x_4 + x_5 = 0.$$

Here $k = 3$, $A = 1$, $B = 2$ so $s = 1$, $t = -2$, and we can take $\lambda_1 = 2$, $\lambda_2 = 0$, $\lambda_3 = 1$ to satisfy $c_1 \lambda_1 + \cdots + c_k \lambda_k = At$. The parameteric solution is then

$$x_1 = a + 2d,$$
$$x_2 = a,$$
$$x_3 = a + d,$$
$$x_4 = d,$$
$$x_5 = d.$$

We now show the "only if" part of Theorem 4. Let $c_1, \ldots, c_n$ be fixed with no subset summing to zero. We shall give a coloring of $Q - \{0\}$ so that (4) has no monochromatic solution.

We introduced a special coloring of $Q - \{0\}$. Let $p > 0$ be prime. Any $q \in Q - \{0\}$ may be uniquely expressed as

$$q = \frac{p^j a}{b}, \qquad j \in Z, a \in Z, b \in N, \gcd(a, b) = 1,$$

$$p \nmid a, p \nmid b.$$

We define $rank(q)$ to be the above-determined j, and we define the *smod* p (smod = super modulo) coloring F_p by

$$F_p(q) = \frac{a}{b} \ (\text{modulo } p) \,. \tag{6}$$

[For example, $F_5(\frac{15}{4}) = \frac{3}{4} = 2$.] F_p is a $(p-1)$-coloring of $Q - \{0\}$. Note that $F_p(x) = F_p(y)$ implies that $F_p(\alpha x) = F_p(\alpha y)$ for all $\alpha \in Q - \{0\}$.

Claim. Assume that p, a prime, does not divide the sum of any nonempty subset of $\{c_i : 1 \leq i \leq n\}$. Then (4) has no monochromatic solutions with the smod p coloring.

The claim clearly implies Theorem 4 since some prime p will not divide any of the (finite number of) *nonzero* sums of $\{c_i\}$.

We assume to the contrary that $x_1, \ldots, x_n$ forms a monochromatic solution to (4). For all $\mu \in Q - \{0\}$, $\mu x_1, \ldots, \mu x_n$ also forms a monochromatic solution. We may thus assume that all $x_i \in Z$, $\gcd(x_1, \ldots, x_n) = 1$. We reorder so that $p \nmid x_i$, $1 \leq i \leq k$; $p \mid x_i$, $k < i \leq n$. Here $k \geq 1$ by the relative primality ($k = n$ is possible). We reduce (4) modulo p:

$$\sum_{i=1}^{n} \bar{c}_i \bar{x}_i = \bar{0} \qquad (\text{in } Z_p) \,,$$

where $\bar{a}$ represents the residue class of a modulo p. Now $\bar{x}_i = \bar{0}$ for $k < i \leq n$ by assumption. Since the x_i are the same color the $\bar{x}_i$, $1 \leq i \leq k$, are equal. Thus

$$\bar{0} = \sum_{i=1}^{n} \bar{c}_i \bar{x}_i = \sum_{i=1}^{k} \bar{c}_i \bar{x}_i = \left(\sum_{i=1}^{k} \bar{c}_i \right) \bar{x}_i \,.$$

Since $\bar{x}_1 \neq \bar{0}$, and p is prime, $\sum_{i=1}^{k} \bar{c}_i = \bar{0}$, contrary to assumptions.

This completes the claim, and therefore Theorem 4 is proved.

3.3 REGULAR HOMOGENEOUS SYSTEMS (RADO'S THEOREM COMPLETE)

We now consider the regularity of systems of linear homogeneous equations. The results are equally complete.

DEFINITION.   A matrix $C = (c_{ij})$ is said to satisfy the *Columns condition* if one can order the column vectors $\mathbf{c}_1, \ldots, \mathbf{c}_n$ and find $1 \leq k_1 < k_2 < \cdots <$

$k_t = n$ such that, setting

$$A_i = \sum_{j=k_{i-1}+1}^{k_i} c_j ,$$

we have

   (i) $A_1 = 0$,

and

   (ii) for $1 < i \leq t$, A_i may be expressed as a rational linear combination of $c_1, \ldots, c_{k_{i-1}}$.

Theorem 5 (Rado's Theorem). The system $Cx = 0$ is regular on N iff C satisfies the Columns condition.

If C has only one row, that is, a single linear homogeneous equation, then Theorem 5 reduces to Theorem 4.

The "only if" section of Rado's theorem involves examining only the smod p colorings. We can state Rado's theorem in an alternative form.

Rado's Theorem (restatement). The system $Cx = 0$ is regular iff for every prime p there is a monochromatic solution under the smod p coloring.

Lemma 6. Let $A, c_1, \ldots, c_k \in Z'$. Suppose that A is not in the vector space (over Q) generated by the c_i. Then, for all but a finite number of primes p, A cannot be expressed as a linear combination of the c_i (modulo p). Moreover Ap^m cannot be expressed as a linear combination of the c_i (modulo p^{m+1}) for any $m \geq 0$.

Proof of Lemma 6. Since A is not in the vector space generated by the c_i we find, by linear algebra, $u \in Q'$ so that $u \cdot \supset_i = 0, 1 \leq i \leq k$, and $u \cdot A \neq 0$. Multiplying u by a suitable constant, we may assume that $u \in Z'$ and $u \cdot A = s \in Z - \{0\}$. Now

$$Ap^m = c_1 x_1 + \cdots + c_k x_k \ (\text{modulo } p^{m+1})$$

implies that

$$sp^m = u \cdot Ap^m = \sum_{i=1}^{k} (u \cdot c_i) x_k = 0 \ (\text{modulo } p^{m+1})$$

so that $p | s$, which holds for only a finite number of primes p.

Fix a matrix C. For every subset $\{c_1, \ldots, c_k\}$ of the column vectors such that $c_1 + \cdots + c_k \neq 0$, let $E(c_1, \ldots, c_k)$ denote the set of (exceptional) primes for which $c_1 + \cdots + c_k = 0$ (modulo p). For every set $\{c_1, \ldots, c_k, A\}$, where the c are column vectors, and $A = c_{k+1} + \cdots + c_r$ for some other column vectors c and A is not a linear combination of the $c_1, \ldots, c_k$, let $E(c_1, \ldots, c_k; A)$ denote the set of primes for which Ap^m is a linear combination of $c_1, \ldots, c_k$ for some m.

Let E denote the union of all $E(c_1, \ldots, c_k)$ and $E(c_1, \ldots, c_k; A)$. E is a finite union of finite sets and therefore finite.

Lemma 7. Fix C. Let E be as defined above. Let p be prime, $p \notin E$. If $Cx = 0$ has a monochromatic solution under the smod p coloring then C satisfies the Columns condition.

Proof of Lemma 7. Let $x_1, \ldots, x_n$ be a monochromatic solution. Reorder by rank (modulo p) so that

$$\begin{aligned} \operatorname{rank}(x_i) = m_1, & \qquad 1 \leq i \leq k_i, \\ m_2, & \qquad k_1 < i \leq k_2, \\ & \qquad \vdots \\ m_t, & \qquad k_{t-1} < i \leq k_t. \end{aligned}$$

Let a be the common color, so that all $x_i = ap^{\operatorname{rank}(x_i)} +$ higher order terms. For convenience we may assume that $m_1 = 0$ by replacing all x_i by $x_i p^{-m_1}$. We write the system $Cx = 0$ as

$$c_1 x_1 + \cdots + c_n x_n = 0.$$

Reduction modulo p gives

$$(c_1 + \cdots + 4c_k)a = 0 \quad (\text{modulo } p)$$

so, as $p \notin E(c_1, \ldots, c_k)$,

$$c_1 + \cdots + c_k = 0.$$

For $1 < j \leq t$, reducing modulo p gives

$$\sum_{i=1}^{k_{j-1}} c_i x_i + Ap^{m_j}a = 0 \quad (\text{modulo } p^{m_j+1}),$$

where

$$A = \sum c_i, \qquad \text{the summation over } k_{j-1} < i \leq k_j.$$

Dividing by a, we find that Ap^{m_j} is a linear combination of $c_1, \ldots, c_{k_{j-1}}$ (modulo p^{m_j+1}) so, since $p \notin E$, A is a linear combiantion of $c_1, \ldots, c_{k_{j-1}}$.

Since this holds for $1 < j \leqslant t$, C satisfies the Columns condition. This completes Lemma 7 and therefore the "only if" section of Rado's theorem.

The "if" section of Rado's theorem requires some preliminaries of interest in their own right.

Theorem 8. Let $G(x_1, \ldots, x_n) = 0$ be a linear homogeneous system of equations that is regular. Let $M > 0$ be fixed. If N is finitely colored there exist $x_1, \ldots, x_n$ satisfying G and $d > 0$ so that all

$$x_i + \lambda d, \qquad 1 \leqslant i \leqslant n, |\lambda| \leqslant M, \tag{7}$$

are the same color.

Proof. Fix the number of colors r. By the Compactness principle there exists R so that any r-coloring of $[R]$ yields a monochromatic solution of system G. Let χ be an r-coloring of N. Define an r^R-coloring χ^* by

$$\chi^*(\alpha) = \chi^*(\beta) \quad \text{iff} \quad \chi(\alpha i) = \chi(\beta i) \quad \text{for } 1 \leqslant i \leqslant R. \tag{8}$$

Set $T = MR^{n-1}$. By van der Waerden's theorem find a monochromatic AP of length $2T + 1$ under χ^*; that is, there exist a and e such that

$$\chi^*(a + \mu e) = \text{constant}, \qquad |\mu| \leqslant T. \tag{9}$$

The r-coloring χ of $a[R]$ yields, by homogeneity, a solution $ay_1, \ldots, ay_n$ ($y_i \in [R]$) with $\chi(ay_i) = \text{constant}$. Now set

$$x_i = ay_i, \qquad 1 \leqslant i \leqslant n,$$
$$d = ey,$$

where $y = lcm(y_1, \ldots, y_n)$. Then, for $|\lambda| \leqslant M$,

$$x_i = \lambda d = ay_i + \lambda ey = y_i\left[a + \lambda e\left(\frac{y}{y_i}\right)\right].$$

Here $|\lambda y / y_i| \leqslant MR^{n-1} = T$ so, by (9),

$$\chi^*\left[a + \lambda e\left(\frac{y}{y_i}\right)\right] = \chi^*(a)$$

and therefore

$$\chi(x_i + \lambda d) = \chi(ay_i) ,$$

which is constant, independent of i.

An example should help to illustrate the beautiful ideas underlying Theorem 8. Suppose that $r = 2$, $R = 10$. We define a $1024 = 2^{10}$-coloring of N, coloring i with the color of $i[10]$. We find an "enormous" AP under this coloring, say

$$S = \{10^9 - 300, \ldots, 10^9 - 3, 10^9, 10^9 + 3, \ldots, 10^9 + 300\} .$$

Now, in the coloring of $[10]$ given by the coloring of $s[10]$ for all $s \in S$, we find a solution, say $y_1 = 2$, $y_2 = 3$, $y_3 = 5$ of G. Then $x_1 = 2 \cdot 10^9$, $x_2 = 3 \cdot 10^9$, $x_3 = 5 \cdot 10^9$ is a monochromatic solution. Each x_i is in the middle of an AP of length 200. Unfortunately the progressions have different periods—6, 9, and 15, respectively. Fortunately they have a common period $lcm(6, 9, 15) = 90$. We set $d = 90$, and since the AP S was "enormous" we can take M "large."

Corollary $8\frac{1}{2}$. Let $G(x_1, \ldots, x_n) = 0$ be a linear homogeneous system of equations. The following are equivalent:

(i) Under any finite coloring of N there exist *distinct* $x_1, \ldots, x_n$ of the same color satisfying $G(x_1, \ldots, x_n) = 0$.
(ii) The system $G(x_1, \ldots, x_n) = 0$ is regular on N, and there exist *distinct* $\lambda_1, \ldots, \lambda_n \in Z$ such that $G(\lambda_1, \ldots, \lambda_n) = 0$.

Proof. Clearly (i) $\Rightarrow$ (ii). Let $\lambda_1, \ldots, \lambda_n$ be *distinct* integers satisfying G. Let $K = \max_{1 \leq i \leq n} |\lambda_i|$. Under any finite coloring of N we find, by Theorem 8, $x_1, \ldots, x_n$ satisfying G and $d > 0$ so that all

$$x_i + \lambda d |\lambda| \leq Kn^2$$

are the same color. For all $\mu, |\mu| \leq n^2$, the values

$$x_i' = x_i + \mu \lambda_i d$$

satisfy $G(x_1', \ldots, x_n') = G(x_1, \ldots, x_n) + \mu d G(\lambda_1, \ldots, \lambda_n) = 0$. If $x_i' = x_j'$ then $\mu = (x_i - x_j)/(\lambda_j - \lambda_i)d$ is determined. For all but at most $\binom{n}{2}$ values of μ, all $x_i' \neq x_j'$. This is the desired distinct solution.

Corollary 9. Let $G(x_1, \ldots, x_n) = 0$ be a linear homogeneous system of equations that is regular. Let $M > 0$ and $c > 0$ be fixed. If N is finitely colored there exist $x_1, \ldots, x_n$ satisfying G and $d > 0$ such that all

$$x_i + \lambda d, \quad 1 \leq i \leq n, |\lambda| \leq M, \tag{10}$$

and

$$cd$$

have the same color.

Proof. Corollary 9 will follow from Theorem 8 in much the same way as Theorem 2 follows from van der Waerden's theorem. We use induction on the number of colors r. We assume that there exists $T = T(r-1, M, s)$ so that if $[T]$ is r-colored there exist $x_1, \ldots, x_n$ satisfying (10). Given an r-coloring of N, we find, by Theorem 8, $x_1, \ldots, x_n$ satisfying G and $d' > 0$ such that all

$$x_i + \lambda d', \quad |\lambda| \leq TM,$$

are the same color. If any $\mu cd'$, $\mu \leq T$, has that color we set $d = \mu d'$ to satisfy (10). Otherwise $cd'[T]$ is $(r-1)$-colored so that (10) is satisfied by induction.

The critical conditions on G in Theorem 8 and Corollary 9 are regularity and homogeneity. Call a family $\mathcal{A}$ of finite subsets of N homogeneous if $A \in \mathcal{A}$, $a \in N$ imply $aA \in \mathcal{A}$, and call $\mathcal{A}$ regular if, whenever N is finitely colored, there exists a monochromatic $A \in \mathcal{A}$.

Corollary 9'. Let $\mathcal{A}$ be homogeneous and regular, $M, c > 0$. If N is finitely colored there exist $A \in \mathcal{A}$, $d > 0$ so that all

$$a + \lambda d, \quad a \in A, |\lambda| \leq M,$$

and

$$cd$$

have the same color.

The proof is identical to that for Corollary 9, replacing "solution to G" by "member of $\mathcal{A}$."

Now we introduce some notation due to W. Deuber.

DEFINITION. $N_{m,p,c} = \{(\lambda_1, \ldots, \lambda_{m+1})$: some $\lambda_i \neq 0$, the first nonzero $\lambda_i = c$, all other $|\lambda_i| \leq p\}$.

A set S of positive integers is called an (m, p, c)-set if

$$S = \left\{ \sum_{i=1}^{m+1} \lambda_i y_i \colon (\lambda_1, \ldots, \lambda_{m+1}) \in N_{m,p,c} \right\}$$

for some $y_1, \ldots, y_m > 0$. For example, $\{x, x + d, x - d, x + 2d, x - 2d, d\}$ is a $(1, 2, 1)$-set.

Theorem 10.   For all $m, p, c > 0$, if N is finitely colored there exists a monochromatic (m, p, c)-set S.

Proof.   We have shown this result for $m = 1$ in Theorem 2. Assume the result for m, p, c so that the family $\mathcal{A}$ of (m, p, c)-sets is regular and, clearly, homogeneous. By Corollary 9' the result now holds for $(m + 1, p, c)$-sets so by induction we are finished.

Completion of Rado's Theorem.   If C has the Columns condition then the system $Cx = 0$ is regular.

Proof.   We show that if C has the Columns condition then the equation $Cx = 0$ has a parametric solution

$$x_i = \lambda_{i1} y_1 + \cdots + \lambda_{in} y_n,$$

where all $\lambda_{ij} \in Z$ and, for each i, the first nonzero λ_{ij} equals c, a constant. As the general case involves cumbersome notation, yet is quite elementary, we shall only illustrate it with an example:

$$
\begin{aligned}
x_1 - x_2 + 3x_3 \qquad\quad + x_5 \qquad &= 0, \\
2x_1 - 2x_2 + 2x_3 + 4x_4 \qquad + x_6 &= 0, \\
3x_1 - 3x_2 + x_3 + 8x_4 + x_5 \qquad &= 0.
\end{aligned}
$$

Here

$$
\begin{aligned}
c_1 &= (1, 2, 3), \\
c_2 &= (-1, -2, -3), \qquad A_1 = c_1 + c_2 = 0, \\
c_3 &= (3, 2, 1), \\
c_4 &= (0, 4, 8) \qquad\qquad\quad A_2 = c_3 + c_4 = 3c_1, \\
c_5 &= (1, 0, 1), \\
c_6 &= (0, 1, 0), \qquad\qquad\quad A_3 = c_5 + c_6 = \tfrac{1}{4}c_1 + \tfrac{1}{4}c_3.
\end{aligned}
$$

Now we may read off

$$(1, 1, 0, 0, 0, 0)$$
$$(-3, 0, 1, 1, 0, 0)$$
$$(-\tfrac{1}{4}, 0, -\tfrac{1}{4}, 0, 1, 1)$$

as rational solutions to $Cx = 0$. We multiply each vector by 4 so as to make all coefficients integral and the "leading" $\lambda_{ij} = 4$. Then

$$
\begin{aligned}
x_1 &= 4y_1 - 12y_2 - y_3 \\
x_2 &= 4y_1 \\
x_3 &= 4y_2 - y_3 \\
x_4 &= 4y_2 \\
x_5 &= 4y_3 \\
x_6 &= 4y_3
\end{aligned}
$$

is a parametric solution of the desired form.

Let C be any matrix satisfying the Columns condition. Let $p = \max|\lambda_{ij}|$; m, c be as above. Any finite coloring of N yields a mono-chromatic (m, p, c)-set that contains a solution to $Cx = 0$.

Much as Rado's dissertation extended Schur's work, the 1973 dissertation of Deuber extended and polished Rado's results. Recall that a system of homogeneous linear equations G is called regular if every finite coloring of N has a monochromatic solution to G. Now call a set $A \subseteq N$ large if every regular system G has a solution in A.

Deuber proves that A is large iff A contains (m, p, c)-sets for all m, p, c. We have already given the main ideas. The condition is necessary since an (m, p, c)-set may be expressed as the solution of a homogeneous G. It is sufficient, as any regular system G may be parameterized so that solutions to G are contained in some (m, p, c)-set.

Deuber goes on to show that, for all m, p, c and r, there exist M, P, C so that an r-coloring of an (M, P, C)-set always contains a mono-chromatic (m, p, c)-set.

Deuber then shows (proving a conjecture of Rado) that the large sets have a surprising partition property. If A is large and $A = A_1 \cup \cdots \cup A_n$ then one of the A_i is large. In particular, if N is finitely colored there exists in one color a solution to all regular equations. These results are not proved in this book.

Deuber has examined the regularity of systems of homogeneous linear equations over arbitrary Abelian groups. (Here we assume that the identity is not colored.) We defer his results to Section 5.4.

3.4 FINITE SUMS AND FINITE UNIONS (FOLKMAN'S THEOREM)

Rado's theorem completely determines the regular systems of homogeneous equations. One special case is of particular interest.

DEFINITION. Let $S \subseteq N$.

$$\mathcal{P}(S) = \left\{ \sum_{s \in S} \varepsilon_s s; \ \varepsilon_s = 0, 1; \ \varepsilon_s = 1 \qquad \text{for a finite nonzero number of } S \right\}.$$

$\mathcal{P}(S)$ is called the *sum-set* of S. For example,

$$\mathcal{P}(\{2, 3, 7\}) = \{2, 3, 5, 7, 9, 10, 12\} .$$

Theorem 11 (Folkman's Theorem). If N is finitely colored there exist arbitrarily large finite sets S such that $\mathcal{P}(S)$ is monochromatic.

Folkman's theorem may be derived as a corollary of Rado's theorem. It is equivalent to the regularity of the system

$$x_T = \sum_{i \in T} x_{\{i\}} , \qquad \emptyset \neq T \subseteq [k] ,$$

which satisfies by elementary, albeit nontrivial, methods the conditions of Rado's theorem. However, the result is of sufficient special interest that we shall give a different proof. Although this result was proved independently by several mathematicians, we choose to honor the memory of our friend Jon Folkman by associating his name with the result.

We shall actually prove Folkman's theorem in the following "finite" form: For a sequence $\{a_i\}$ and a finite nonempty set I, let $a(I)$ denote $\sum_{i \in I} a_i$.

Folkman's Theorem (restatement).   For all c and there exists $M = M(c, k)$ so that, if $[M]$ is c-colored, there exist $a_1, \ldots, a_k$ so that all $a(I)$ are colored identically.

The following critical lemma is based on van der Waerden's theorem. Let $W(c, k)$ denote van der Waerden's function, where c is the number of colors and k is the desired length of the AP.

Lemma 12. For all c, k there exists $n = n(c, k)$ so that, if $[n]$ is c-colored, there exist $a_1 < a_2 < \cdots < a_k$ with all $a(I) \leq n$ so that the color of $a(I)$ depends only on $\max(I)$.

Proof. We prove the existence of $n(c, k)$ for all c by induction on k. For $k = 1$ (even $k = 2$) it is trivial. We claim that we may take $n = n(c, k + 1) = 2W(c, n(c, k))$. Let a c-coloring of $[n]$ be given. By examining only $\{n/2 + 1, \ldots, n\}$ we find a_{k+1}, d with $n(c, k) < a_{k+1}$ (actually $n/2 < a_{k+1}$) so that

$$\{a_{k+1} + \lambda d : 0 \le \lambda \le n(c, k)\}$$

is monochromatic, say, red. Now, identifying $d[n(c, k)]$ with $[n(c, k)]$, we can find $a_1 < \cdots < a_k$, all a_i divisible by d and their sum at most $dn(c, k)$, so that $\{a_1, \ldots, a_k\}$ satisfies the induction hypothesis. Consider $A = \{a_1, \ldots, a_{k+1}\}$. For $j < k + 1$ the $a(I)$ where $\max(I) = j$ are monochromatic by the induction hypothesis. If $\max(I) = k + 1$ then $a(I) = a_{k+1} + \lambda d$, where $0 \le \lambda \le n(c, k)$, so that $a(I)$ is red. Thus A satisfies the induction hypothesis for $k + 1$, completing the induction.

Proof of Folkman's Theorem. Our lemma allows us to use the Induced Color method. We take $M = M(c, k) = n(c, (c - 1)k + 1)$. Given a c-coloring on $[M]$, we find $a_1 < \cdots < a_{(c-1)k+1}$, satisfying the lemma. Now we define a coloring on $[(c - 1)k + 1]$ by coloring i with the color of all $a(I)$, with $\max(I) = i$. By the Pigeon-Hole principles we find a subset $S \subseteq [(c - 1)k + 1]$, $|S| = k$, monochromatic under the induced coloring. We set $A = \{a_i : i \in S\}$. Then $\mathcal{P}(A)$ is monochromatic.

Folkman's theorem has an analogue in set theory, with set union taking the place of sum. Call a family of sets $\mathcal{D}$ a *disjoint* collection if the elements of $\mathcal{D}$ are pairwise disjoint finite sets. Write $\mathcal{D} = \{D_i\}_{i \in I}$, where I is a finite indexing set. Let $FU(\mathcal{D})$ denote the family of all finite unions of the $D \in \mathcal{D}$, that is,

$$FU(\mathcal{D}) = \left\{ \bigcup_{i \in T} D_i : \emptyset \ne T \subseteq I, T \text{ finite} \right\}.$$

Let $\mathcal{P}(X)$ denote the family of nonempty finite subsets of X, and let $\mathcal{P}_n$ denote $\mathcal{P}([n])$.

Theorem 13 (Finite Unions Theorem). If the finite subsets of N are finitely colored there exist arbitrarily large $\mathcal{D}$ so that $FU(\mathcal{D})$ is monochromatic. Again, we shall prove the finite form.

Finite Unions Theorem. For all k, c there exists $F = F(k, c)$ such that, if $n \ge F$ and $\mathcal{P}_n$ is c-colored, there exists a disjoint collection $\mathcal{D}$ of cardinality k such that $FU(\mathcal{D})$ is monochromatic.

We shall outline two proofs of the Finite Unions theorem. We first show that Folkman's theorem and the Finite Unions theorems are equivalent in the imprecise sense that each can be quickly deduced from the other. There is a natural correspondence between N and $\mathcal{P}(N)$, given by

$$\varphi(I) = \sum_{i \in I} 2^{i+1}.$$

Assuming the Finite Unions theorem, we show that $M(k, c) \leq 2^{F(k,c)}$. A c-coloring of $[2^{F(k,c)}]$ corresponds under φ^{-1} to a c-coloring of $\mathcal{P}_{F(k,c)}$ in which there is a disjoint collection $\mathcal{D}$ of cardinality k. For which $FU(\mathcal{D})$ is monochromatic. But union of disjoint sets corresponds, under φ, to addition of integers so that, in the original coloring, all finite sums of $\varphi(\mathcal{D})$ are the same color.

The converse is less obvious. Assume Folkman's theorem, and fix k, c. Select F so large (by Chapter 1, Theorem 10) that if $n \geq F$ and $\mathcal{P}_n$ is c-colored there exists $B \subseteq [n]$, $|B| = M(k, c)$, where, for $1 \leq i \leq M(k, c)$, $[B]^i$ is monochromatic. For such F and a coloring of $\mathcal{P}_n$, $n \geq F$, we find B as above. We define a coloring on $[M(k, c)]$ by giving i the color of all $X \in [B]^i$. In the induced coloring we find $a_1, \ldots, a_k$ so that all finite sums are monochromatic. Now we simply set $\mathcal{D} = \{D_1, \ldots, D_k\}$, where the D_i are pairwise disjoint subsets of B with $|D_i| = a_i$. Any finite union of the D_i is a subset of B so that its cardinality determines its color. But the cardinality of a finite union of disjoint sets is just the finite sum of the cardinalities so that, indeed, $FU(\mathcal{D})$ is monochromatic.

A second proof of the Finite Unions theorem is based on the Hales–Jewett theorem. The vertices of the n-cube $[0, 1]^n$ may be placed in a natural correspondence with $\mathcal{P}_n$. We then interpret the Extended Hales–Jewett theorem (Chapters 2, Theorem 7) as follows: For all k, c there exists n so that, if the subsets of $[n]$ (including the null set) are c-colored, there exist disjoint $A_0, A_1, \ldots, A_k$ so that all

$$A_0 \cup \bigcup_{i \in I} A_i, \qquad I \subseteq \{1, \ldots, k\}$$

are monochromatic. If A_0 were the null set we would be finished. However, the result obtained above bears exactly the same relation to the Finite Unions theorem as van der Waerden's theorem does to Folkman's theorem, and the proof follows exactly the same lines. We omit the details.

We may replace adding by multiplication in Folkman's theorem.

Define

$$\mathscr{P}'(S) = \left\{ \prod_{s \in S} s^{\varepsilon_s}, \quad \varepsilon_s = 0, 1; \varepsilon_s = 1 \quad \begin{array}{l} \text{for a finite nonzero} \\ \text{number of } s \end{array} \right\}$$

For example, $\mathscr{P}'(\{2, 3, 7\}) = \{2, 3, 6, 7, 14, 21, 42\}$, the set of finite products.

Theorem 14. If N is finitely colored there exist arbitrarily large S such that $\mathscr{P}'(S)$ is monochromatic.

Proof. We need only examine the coloring of $\{2^n : n \geq 1\}$. Here multiplication mirrors addition (recall that ancient instrument—the slide rule) so this result is a corollary of Folkman's theorem.

Conjecture. If N is finitely colored there exist arbitrarily large S so that $\mathscr{P}(S) \cup \mathscr{P}'(S)$ is monochromatic.

This conjecture has proved surprisingly intractable. Even for $|S| = 2$ it is an open question whether, if N is finitely colored, there must exist a monochromatic $\{x, y, x + y, xy\}$. N. Hindman has given a 2-coloring of N for which no infinite S exists with $\mathscr{P}(S) \cup \mathscr{P}'(S)$ monochromatic.

3.5 INFINITE SETS OF SUMS (HINDMAN'S THEOREM)

It was natural to ask, and was conjectured for some time, whether Folkman's theorem could be extended to infinite sets S. An affirmative answer is given in the next theorem. Although this infinite result is technically beyond the scope we have set for this book, we believe that the result and proof are so interesting as to warrant this exception.

Theorem 15 (Hindman's Theorem). If N is finitely colored there exists $S \subseteq N$ S *infinite*, such that $\mathscr{P}(S)$ is monochromatic.

We emphasize that Hindman's theorem is *not* a corollary of Folkman's theorem. Compactness does not work "in reverse"; the existence of finite arbitrarily large monochromatic structures does not imply the existence of infinite monochromatic structures. For example, if N is finitely colored there does not necessarily exist an infinite monochromatic arithmetic progression.

This result is due to Hindman. The proof was greatly simplified, though the same basic ideas were used, by J. Baumgartner, and it is his proof we present. In Sections 6.1 and 6.2 we give alternative proofs involving noncombinatorial methods.

In proving Theorem 15, Baumgartner considers a set-theoretic Ramsey theorem. We shall change our notation slightly from the finite results. Call $\mathcal{D}$ a disjoint collection if $\mathcal{D}$ is an *infinite* collection of disjoint finite sets, and let $FU(\mathcal{D})$ denote the family of all finite unions of elements of $\mathcal{D}$ (excluding the empty union).

Theorem 16. Let $[N]^{<\omega} = \mathcal{C}_1 + \cdots + \mathcal{C}_k$. Then there exist $1 \le i \le k$ and a disjoint collection $\mathcal{D}$ with

$$\mathcal{C}_i \supseteq FU(\mathcal{D}).$$

Theorem 16 implies Hindman's theorem by a use of the canonical bijection between N and $[N]^{<\omega}$, letting $n = \Sigma \, \varepsilon_i 2^i$ correspond with $\{i: \varepsilon_i = 1\}$. The proof of Theorem 16 will require a sequence of lemmas.

On the class of disjoint collections we define a partial order $<$ by $\mathcal{D}_1 < \mathcal{D}$ iff $\mathcal{D}_1 \subseteq FU(\mathcal{D})$. The crucial definition is that $\mathcal{C}$ is *large for* $\mathcal{D}$ if $\mathcal{C} \cap FU(\mathcal{D}_1) \ne \varnothing$ for all $\mathcal{D}_1 < \mathcal{D}$.

Remarks

$\mathcal{D}_1 < \mathcal{D}$ implies $FU(\mathcal{D}_1) \subseteq FU(\mathcal{D})$.

$\mathcal{C}$ large for $\mathcal{D}$ and $\mathcal{D}_1 < \mathcal{D}$ imply $\mathcal{C}$ large for $\mathcal{D}_1$

$FU(\mathcal{D})$ is large for $\mathcal{D}$. In particular, $[N]^{<\omega}$ is large for $[N]^1$.

$\mathcal{C}$ is large for $\mathcal{D}$ iff $\mathcal{C} \cap FU(\mathcal{D})$ is large for $\mathcal{D}$.

$\mathcal{C}$ large for $\mathcal{D}$ and $\mathcal{C} \subseteq \mathcal{C}'$ imply $\mathcal{C}'$ large for $\mathcal{D}$.

Lemma 17 (Decomposition Lemma). Assume that $\mathcal{C}$ is large for $\mathcal{D}$ and $\mathcal{C} = \mathcal{C}_1 + \cdots + \mathcal{C}_k$. Then there exist $1 \le i \le k$ and $\mathcal{D}_1 < \mathcal{D}$ so that $\mathcal{C}_i$ is large for $\mathcal{D}_1$.

Proof. Let $k = 2$. If $\mathcal{C}_1$ is not large for $\mathcal{D}$ then $\mathcal{C}_1 \cap FU(\mathcal{D}_1) = \varnothing$ for some $\mathcal{D}_1 < \mathcal{D}$. For any $\mathcal{D}_2 < \mathcal{D}_1$, $\mathcal{C} \cap FU(\mathcal{D}_2) \ne \varnothing$ so that $\mathcal{C}_2 \cap FU(\mathcal{D}_2) \ne \varnothing$ and hence $\mathcal{C}_2$ is large for $\mathcal{D}_1$.

The general case follows by induction.

Theorem 18. If $\mathcal{C}$ is large for $\mathcal{D}$ there exists $\mathcal{D}_1^* < \mathcal{D}$ so that $\mathcal{C} \supseteq FU(\mathcal{D}_1^*)$.

The proof requires a series of lemma. We first show that Theorem 18 implies Theorem 16. Let $[N]^{<\omega} = \mathcal{C}_1 + \cdots + \mathcal{C}_k$. Since $[N]^{<\omega}$ is large for $[N]^1$ there exists $\mathcal{D} < [N]^1$ and i so that $\mathcal{C}_i$ is large for $\mathcal{D}$. Theorem 18 then implies $\mathcal{C}_i \supseteq FU(\mathcal{D}_1)$ for some $\mathcal{D}_1 < \mathcal{D}$. Define

$$\mathcal{C} - S = \{C \in \mathcal{C}: C \cap S = \varnothing\}.$$

Lemma 19. $\mathscr{C}$ large for $\mathscr{D}$ and S finite imply $\mathscr{C} - S$ large for $\mathscr{D}$.

Proof. Suppose that $\mathscr{D}_1 < \mathscr{D}$ with $(\mathscr{C} - S) \cap FU(\mathscr{D}_1) = \varnothing$. Let $\mathscr{D}_2 = \{D \in \mathscr{D}_1 : D \cap S = \varnothing\}$. ($\mathscr{D}_2$ is infinite since S is finite.) Then $\mathscr{C} \cap FU(\mathscr{D}_2) = \varnothing$, but $\mathscr{D}_2 < \mathscr{D}$, and we reach a contradiction.

Lemma 20. Assume $\mathscr{C}$ large for $\mathscr{D}$. There exists $S \in FU(\mathscr{D})$, $\mathscr{D}_1 < \mathscr{D} - S$, so that

$$\mathscr{C}_1 = \{T \in \mathscr{C} : T \cap S = \varnothing, T \cup S \in \mathscr{C}\} \text{ is large for } \mathscr{D}_1 .$$

Proof. There must exist $n, D_1, \ldots, D_n \in \mathscr{D}$ disjoint so that, for every $D_{n+1} \in FU(\mathscr{D})$ disjoint from $D_1 \cup \cdots \cup D_n$, some $D_{n+1} \cup D_I \in \mathscr{C}$ (where $I \subseteq \{1, \ldots, n\}$, $I \neq \varnothing$, and we define $D_I = \cup_{i \in I} D_i$). Otherwise we could construct $\mathscr{D}' = \{D_1, D_2, \ldots\}$ so that $\mathscr{C} \cap FU(\mathscr{D}') = \varnothing$ and never get "stuck." We fix $n, D_1, \ldots, D_n, D^* = D_1 \cup \cdots \cup D_n$. For $\varnothing \neq I \subseteq \{1, \ldots, n\}$ we define

$$\mathscr{C}_I = \{C \in \mathscr{C} : C \cap D^* \neq \varnothing, C \cup D_I \in \mathscr{C}\} .$$

The $\mathscr{C}_I$ give a finite decomposition of $\mathscr{C} - D^*$, and $\mathscr{C} - D^*$ is large for $\mathscr{D} - D^*$ so some $\mathscr{C}_I$ is large for some $\mathscr{D}_I < \mathscr{D} - D^* < \mathscr{D} - D_I$, implying the lemma with $S = D_I$.

Lemma 21. Assume that $\mathscr{C}$ is large for $\mathscr{D}$. There exists $S' \in \mathscr{C} \cap FU(\mathscr{D})$, $\mathscr{D}' < \mathscr{D}$, so that

$$\mathscr{C}' = \{T \in \mathscr{C} : T \cap S' = \varnothing, T \cup S' \in \mathscr{C}\} \text{ is large for } \mathscr{D}' .$$

Proof. The requirement $S' \in \mathscr{C}$ distinguishes Lemmas 21 and 20. We apply Lemma 20 repeatedly. Beginning with $\mathscr{C}_0 = \mathscr{C}$, $\mathscr{D}_0 = \mathscr{D}$, we find, for $i \geq 1$, $S_i, \mathscr{C}_i, \mathscr{D}_i$ with $S_{i+1} \in FU(\mathscr{D}_i)$ so that

$$\mathscr{C}_{i+1} = \{T \in \mathscr{C}_i : T \cap S_{i+1} = \varnothing, T \cup S_{i+1} \in \mathscr{C}_i\}$$

is large for $\mathscr{D}_{i+1} < \mathscr{D}_i$ and $D \cap \cup_{j=1}^{i} S_j = \varnothing$ for all $D \in FU(\mathscr{D}_{i+1})$. The S_i form a disjoint collection so we find $i_1 < \cdots < i_k$,

$$S' = S_{i_1} \cup \cdots \cup S_{i_k} \in \mathscr{C}$$

Now, if $T \in \mathscr{C}_{i_k}$, $T \cup S' \in \mathscr{C}$ (by the definition of the $\mathscr{C}_i$, $T \cup S \in \mathscr{C}$ for all partial unions S of the $S_{i_1}, \ldots, S_{i_k}$), and Lemma 21 holds with $\mathscr{D}' = \mathscr{D}_{i_k}$ as $\mathscr{C}' \supseteq \mathscr{C}_{i_k}$.

Proof of Theorem 18 (and hence Theorems 15 and 16). By repeated applications of Lemma 21 we find S^i, $\mathscr{C}^i$, $\mathscr{D}^i$ so that $S^{i+1} \in \mathscr{C}^i \cap FU(\mathscr{D}^i)$ and

$$\mathscr{C}^{i+1} = \{T \in \mathscr{C}^i : T \cap S^{i+1} = \varnothing, T \cup S^{i+1} \in \mathscr{C}^i\} \quad \text{is large for } \mathscr{D}^i.$$

Then $\mathscr{D}^* = \{S^1, S^2, \ldots\}$ is the desired set.

3.6 REGULAR NONHOMOGENEOUS SYSTEMS

The results on nonhomogeneous linear systems are far simpler than those for the homogeneous case. We express our results in a fashion that will be particularly appropriate for Section 5.6. We restrict our attention to a single equation; the straightforward generalization to systems can be found in Rado's original paper.

Lemma 22. There is a $(2n)$-coloring χ of Q so that

$$\sum_{i=1}^{n} (y_i - y_i') = 1 \tag{11}$$

has no solution with $\chi(y_i) = \chi(y_i')$, $1 \leqslant i \leqslant n$.

Proof. Define χ by setting, for $0 \leqslant j \leqslant 2n - 1$,

$$\chi(y) = j \quad \text{iff} \quad y \in \left[2m + \frac{j}{n}, 2m + \frac{j+1}{n}\right) \quad \text{for some } m \in Z. \tag{12}$$

Then $\chi(y_i) = \chi(y_i')$ implies that $y_i - y_i' = 2m_i + \Theta_i$, $|\Theta_i| < n^{-1}$, so

$$\sum_{i=1}^{n} (y_i - y_i') = 2 \sum_{i=1}^{n} m_i + \Theta,$$

where

$$\Theta = \sum_{i=1}^{n} \Theta_i \quad \text{and} \quad |\Theta| < \sum_{i=1}^{n} |\Theta_i| < 1.$$

Theorem 23. Let Ω be any field of characteristic zero, $c_1, \ldots, c_n, b \in \Omega$, $b \neq 0$. There is a $(2n)^n$-coloring χ^* of Ω so that

$$\sum_{i=1}^{n} c_i(x_i - x_i') = b \tag{13}$$

has no solution with $\chi^*(x_i) = \chi^*(x_i')$, $1 \leqslant i \leqslant n$.

Proof. Considering Ω as a vector space over Q, we may find a linear mapping

$$\psi : \Omega \to Q ,$$

$$\psi(b) = 1 .$$

Define χ by (12) and χ^* by

$$\chi^*(\alpha) = \chi^*(\beta) \quad \text{iff} \quad \chi(\psi(c_i \alpha)) = \chi(\psi(c_i \beta)) \quad \text{for } 1 \le i \le n .$$

Then χ^* is a $(2n)^n$-coloring of Ω. If (13) holds with $\chi^*(x_i) = \chi^*(x_i')$, $1 \le i \le n$, then

$$\sum_{i=1}^{n} [\psi(c_i x_i) - \psi(c_i x_i')] = \psi(b) = 1 ,$$

and $\chi(\psi(c_i x_i)) = \chi(\psi(c_i x_i'))$, $1 \le i \le n$, contradicting Lemma 20.

The proof of Theorem 23 involves the Axiom of Choice (to find ψ). For $\Omega = R$, Lemma 22 may be extended directly.

Corollary 24.   Let Ω be a field of characteristic zero. The equation

$$c_0 x_0 + c_1 x_1 + \cdots + c_n x_n = b , \qquad c_i, b \in \Omega, b \ne 0 , \tag{14}$$

is regular on Ω iff $\Sigma_{i=0}^{n} c_i \ne 0$.

Proof.   If $\Sigma_{i=0}^{n} c_i = A \ne 0$ then $x_1 = \cdots = x_n = b/A$ is always a monochromatic solution to (14). If $\Sigma_{i=0}^{n} c_i = 0$ then (14) becomes $\Sigma_{i=1}^{n} c_i(x_i - x_0) = b$ so that there is no monochromatic solution under the $(2n)^n$-coloring χ^* of Theorem 23.

REMARKS AND REFERENCES

§1.   Schur [1916] provides the original reference. Mirsky [1975] gives a tribute to Schur and an overview of the work on Schur's theorem.

§2, 3, 6.   See Rado [1943] and also Rado [1933a], [1933b], [1936], and [1969], as well as Deuber [1973].

§4.   Independent proofs of Folkman's theorem are given by Sanders [1969] and Rado [1969].

§5.   Hindman [1974] and Baumgartner [1974] present proofs of Hinderman's theorem. Milliken [1975] gives an interesting generalization.

§6.   Straus [1975] extends these results to arbitrary Abelian groups.

4

Numbers

Most of the results of Chapters 1–3 state that an r-coloring of any sufficiently large structure contains a monochromatic substructure of a certain size. In this chapter we concern ourselves with precisely how large such a structure need be. To the existential results of the preceding chapters we associate functions. Evaluation of these functions has proved to be extremely difficult. Our best results, for Ramsey's theorem itself, are still far from the original expectations.

4.1 RAMSEY NUMBERS—EXACT

A prodigious amount of effort has gone into finding the exact values of the Ramsey function $R(k, l)$ for small values of k, l. [The Ramsey functions are defined in Section 1.1. In graph-theoretic terms $R(k, l)$ is the minimal n so that any graph on n vertices contains either a clique of size k or an independent set of size l.] In 1955, R. E. Greenwood and A. M. Gleason found the values for $(k, l) = (3, 3), (3, 4), (3, 5), (4, 4)$. [Trivially, R is symmetric and $R(k, 2) = k$.] Since then, only two other exact values have been found. Table 4.1 gives all known exact bounds and some upper and lower bounds on the function R. It is unlikely that substantial improvement will be made on this table. See cite (RAD) for the most current information. Even evaluation of $R(5, 5)$ appears beyond current man-machine capabilities.

Proof 1 of Ramsey's theorem—abridged (Chapter 1, Theorem 1) gives

$$R(k, l) \leq R(k, l-1) + R(k-1, l). \tag{1}$$

A close examination reveals a slight improvement. Let $n = R(k, l-1) + R(k-1, l) - 1$. If $[n]^2$ is 2-colored with neither a red K_k nor a blue K_l then each point x is connected to the remaining $n - 1$ points by precisely $R(k-1, l) - 1$ red lines and $R(k, l-1) - 1$ blue lines. Hence the total number of red lines is exactly $n(R(k-1, l) - 1)/2$, which must be an

Table 4.1   The Ramsey Function $R(k, l)$

k \ l	3	4	5	6	7	8	9
3	6	9	14	18	23	28/29	36
4		18	25/28	34/44			
5			43/55	51/94			
6				102/178			

integer. This is impossible if $R(k-1, l)$ and $R(k, l-1)$ are even. Thus, in that case, inequality (1) is strict.

The foregoing arguments are sufficient to give the upper bounds for $R(3,3)$, $R(3,4)$, $R(3,5)$, and $R(4,4)$. More precise techniques are required, however, for larger values. A lower bound $R(k, l) > n$ requires the construction of a graph on n vertices containing neither k-clique nor l-independent set. For these four values the graphs are given in Fig. 4.1.

These graphs have considerable structure. In Fig. 4.1d, the vertices are Z_{17} and $\{i, j\}$ is an edge iff $i - j$ is a square in Z_{17}. Figure 4.1a is defined identically over Z_5. In Fig. 4.1c, the vertices are Z_{13} and $\{i, j\}$ is an edge iff $i - j$ is a cubic residue. Figure 4.1b consists of the vertices of Fig. 4.1c not adjacent to 0.

These results were all known to Greenwood and Gleason. Many unsuccessful efforts were made to extend them. It appears likely (though not certain) that the structure of these maximal Ramsey graphs is illusory. Perhaps combinatorialists have again been victimized by the Law of Small Numbers: Patterns discovered for small k evaporate for k sufficiently large to make calculation difficult.

When the number of colors is arbitrary, the proof of Ramsey's theorem gives

$$R(k_1, \ldots, k_s) \leqslant 2 + \sum_{i=1}^{s} [R(k_1, \ldots, k_{i-1}, k_i - 1, k_{i+1}, \ldots, k_s) - 1]. \quad (2)$$

Since $R(k_1, k_2, 2) = R(k_1, k_2)$, this implies that $R(3,3,3) \leqslant 17$. Greenwood and Gleason define a 3-coloring of K_{16}, labeling the vertices by $GF(16)$ and coloring $\{\alpha, \beta\}$ by the cubic character of $\alpha - \beta$. They prove that there are no monochromatic triangles; hence $R(3,3,3) = 17$. This is the only nontrivial Ramsey number known for more than two colors.

In Section 1.2 we defined R_s, the Ramsey function for coloring s-tuples. For $s > 2$ no exact values of R_s are known. The first nontrivial case is $R_3(4)$: the minimal n so that, given any 2-coloring of $[n]^3$, there exists a four-element set all of whose three-element subsets are the same color. The best bounds as of this writing are $13 \leqslant R_3(4) \leqslant 15$.

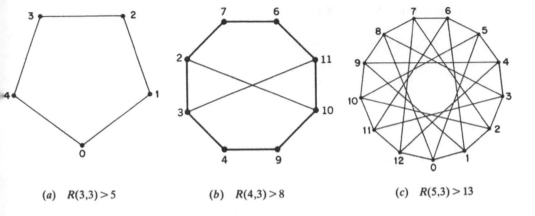

(a) $R(3,3) > 5$ (b) $R(4,3) > 8$ (c) $R(5,3) > 13$

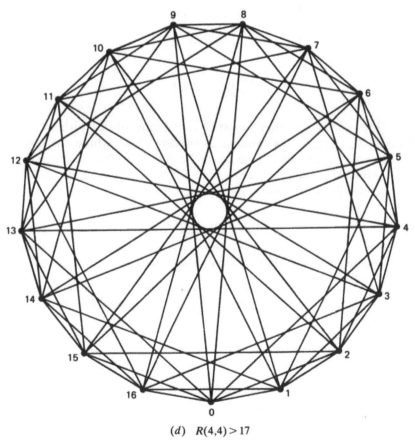

(d) $R(4,4) > 17$

Figure 4.1 Small Ramsey graphs.

4.2 RAMSEY NUMBERS—ASYMPTOTICS

From recursion (1) we may derive

$$R(k, l) \leq \binom{k+l-2}{k-1} \tag{3}$$

so that $R(k) \leq c4^k k^{-1/2}$. The second proof of Ramsey's theorem gives only $R(k) \leq c4^k$. (In this section c denotes an appropriate constant.) For the lower bound we use an *Existence argument*. This method of proof, also called the probabilistic or nonconstructive method, enables one to prove the existence of finite structures having certain properties without actually constructing the structures themselves.

Theorem 1. $R(k) > k2^{k/2}[(1/e\sqrt{2}) + o(1)]$.

Proof. More precisely, we show that if

$$\binom{n}{k} 2^{1-\binom{k}{2}} < 1 \tag{4}$$

then $R(k) > n$; that is, there exists a 2-coloring of K_n without monochromatic K_k. Consider a *random* 2-coloring of K_n where the color of each edge is determined by the toss of a fair coin. More precisely, we have a probability space whose elements are the 2-colorings of K_n and whose probabilities are determined by setting

$$P[\{i, j\} \text{ is red}] = \tfrac{1}{2} \tag{5}$$

for all i, j and making these probabilities mutually independent. Thus there are $2^{\binom{n}{2}}$ colorings, each with probability $2^{-\binom{n}{2}}$. For any set of vertices S, $|S| = k$, let A_S denote the event "S is monochromatic." Then

$$P[A_S] = 2^{1-\binom{k}{2}}, \tag{6}$$

as the $\binom{k}{2}$ "coin flips" to determine the colors of $[S]^2$ must be the same.

The event "some k-element set of vertices S is monochromatic" is represented by $\bigvee_{|S|=k} A_S$:

$$P\left[\bigvee_{|S|=k} A_S\right] \leq \sum_{|S|=k} P[A_S]$$

$$\leq \binom{n}{k} 2^{1-\binom{k}{2}} < 1 \qquad (7)$$

under assumption (4). Thus *some* coloring is in the complement of this event. That is the desired coloring.

We have shown that

$$\sqrt{2} \leq \liminf R(k, k)^{1/k} \leq \limsup R(k, k)^{1/k} \leq 4 . \qquad (8)$$

The value of $\lim R(k, k)^{1/k}$ is not known and is the major open problem involving the asymptotics of the Ramsey function. (Even the existence of the limit is not known.) Another problem is the nonconstructive nature of the Existence argument. It would be of great interest to construct ("construct" is not precisely defined) a 2-coloring of K_n for n large containing no monochromatic K_k. For example, as mentioned previously, if $n = 4t + 1$ is prime one can 2-color Z_n by coloring $\{i, j\}$ by the quadratic character of $i - j$. Although this appears to give good results for small n, the number-theoretic problems raised by asymptotic considerations seem (at present) unresolvable.

Theorem 2. If, for some $p, 0 \leq p \leq 1$,

$$\binom{n}{k} p^{\binom{k}{2}} + \binom{n}{l}(1 - p)^{\binom{l}{2}} < 1 \qquad (9)$$

then $R(k, l) > n$.

Proof. We use the Existence argument of Theorem 1, replacing (5) by

$$P[\{i, j\} \text{ is red}] = p . \qquad (10)$$

For $S, |S| = k$ let A_S be the event "$[S]^2$ is red," and for $T, |T| = l$ let B_T be the event "$[T]^2$ is blue." Then

$$P\left[\bigvee_S A_S \vee \bigvee_T B_T\right] < 1 \qquad (11)$$

so the desired coloring of K_n exists.

The following probability result, due to L. Lovász, fundamentally improves the Existence argument in many instances. Let $A_1, \ldots, A_n$ be events in a probability space Ω. A graph G on $[n]$ is said to be a dependency graph of $\{A_i\}$ if, for all i, the event A_i is *mutually independent* of $\{A_j: \{i, j\} \notin E(G)]\}$. A_i must be not only independent of each A_j but of any combination of the A_j.

Theorem 3 (Lovász Local Lemma). Let $A_1, \ldots, A_n$ be events with a dependency graph G. Suppose that there exists $x_1, \ldots, x_n, 0 < x_i < 1$, so that, for all i,

$$P(A_i) < x_i \prod_{\{i,j\} \in E(G)} (1 - x_j). \tag{12}$$

Then $P(\bigwedge \bar{A}_i) > 0$.

Proof. We show that

$$P\left(A_i \Big| \bigwedge_S \bar{A}_j\right) < x_i \tag{13}$$

for all i and S with $i \notin S$. If $S = \emptyset$, (13) follows directly from (12). We use induction on $|S|$. Fix i, S. Let

$$U = \{j: \{i, j\} \in E(G)\},$$
$$T = S \cap U.$$

Renumber so that $T = \{1, \ldots, t\}$. Then

$$P\left(A_i \Big| \bigwedge_S \bar{A}_j\right) = \frac{P\left(A_i \wedge \bigwedge_T \bar{A}_j \Big| \bigwedge_{S-T} \bar{A}_j\right)}{P\left(\bigwedge_T \bar{A}_j \Big| \bigwedge_{S-T} \bar{A}_j\right)} \tag{14}$$

by the general equality $P(A|BC) = P(AB|C)/P(B|C)$. We bound

$$P\left(A_i \wedge \bigwedge_T \bar{A}_j \Big| \bigwedge_{S-T} \bar{A}_j\right) \le P\left(A_i \Big| \bigwedge_{S-T} \bar{A}_j\right) = P(A_i) \tag{15}$$

by the assumption of independence. The denominator satisfies

$$P\Big(\bar{A}_1 \ldots \bar{A}_t \big| \bigwedge_{S-T} \bar{A}_j\Big) = \prod_{r=1}^{t} P\Big(\bar{A}_r \big| \bar{A}_{r+1} \ldots \bar{A}_t \wedge \bigwedge_{S-T} \bar{A}_j\Big)$$

$$> \prod_{r \in T} (1 - x_r) \tag{16}$$

by the induction assumption (13). Combining the two bounds gives

$$P\Big(A_i \big| \bigwedge_{S} \bar{A}_j\Big) < \frac{P(A_i)}{\prod_{r \in T} (1 - x_r)} < x_i \prod_{r \in U-T} (1 - x_r) \leqslant x_i, \tag{17}$$

completing the induction. Finally,

$$P(\bar{A}_1 \ldots \bar{A}_n) = \prod_{i=1}^{n} P(\bar{A}_i | \bar{A}_1 \ldots \bar{A}_{i-1}) > 0. \tag{18}$$

A special case of particular interest occurs when the A_i are symmetric in some sense and all x_i are chosen to be equal.

Corollary 4. Let $A_1, \ldots, A_n$ be events with $P(A_i) \leqslant p$ for all i and with a dependency graph G of maximal degree at most d, that is, for all i,

$$|\{j: \{i, j\} \in E(G)\}| \leqslant d.$$

If

$$ep(d + 1) < 1 \tag{19}$$

then $P(\bigwedge \bar{A}_i) > 0$.

Proof. We apply Theorem 3 with $x_1 = \cdots = x_n = 1/(d + 1)$. Condition (12) then becomes

$$p < \frac{d^d}{(d + 1)^{d+1}},$$

which we have weakened slightly to facilitate applications.

We apply this method to the proof of Theorem 1. We define a dependency graph on the events A_S, defined in the proof by joining S to T if $|S \cap T| \geqslant 2$. The dependency graph is regular, and

$$d = |\{T : |T \cap S| \geq 2\}| \leq \binom{k}{2}\binom{n}{k-2},$$

$$p = 2^{1-\binom{k}{2}}.$$

If $ep(d+1) < 1$ then $R(k) > n$. Thus (after an asymptotic analysis)

$$R(k) > k 2^{k/2}\left[\frac{\sqrt{2}}{e} + o(1)\right].\qquad(20)$$

The best asymptotic bounds on $R(k, l)$ are obtained in the case $l = 3$.

Theorem 5. $ck^2/\log k \leq R(3, k) \leq c'k^2/\log k$.

We do not prove Theorem 5 in this book. References may be found at the end of this chapter.

For fixed $l > 3$ it is conjectured that $R(k, l) = k^{l-1+o(1)}$, asymptotically in k.

4.3   VAN DER WAERDEN NUMBERS

Recall that $W(k, t)$ denotes the minimal integer so that if $[W]$ is t-colored there exists a monochromatic arithmetic progression of k-terms. The nontrivial exact values of W known are $W(3, 2) = 9$, $W(4, 2) = 35$, $W(3, 3) = 27$, $W(3, 4) = 76$, $W(5, 2) = 178$.

Let $W(k) = W(k, 2)$. The best known upper bound for $W(k)$ is a wowzer function (see Section 2.7) given by adapting the Shelah proof of Section 2.6 to van der Waerden's theorem. Nearly identical asymptotic lower bounds for $W(k)$ are achieved by the following two theorems, with completely different methods of proof. Perhaps $W(k)$ is actually of this order of magnitude—though this is scant evidence on which to base a conjecture!

Theorem 6. If p is prime, $W(p+1) \geq p2^p$.

Proof. We show only the slightly weaker result $W(p+1) \geq p(2^p - 1)$. Let $GF(2^p)$ denote the finite field with 2^p elements, and fix $\alpha \in GF(2^p)$, α primitive [i.e., α generates the cyclic multiplicative group $GF(2^p)^*$]. Fix a basis $v_1, v_2, \ldots, v_p$ for $GF(2^p)$ over Z_2. For any integer j set

$$\alpha^j = a_{1j}v_1 + a_{2j}v_2 + \cdots + a_{pj}v_p, \qquad a_{ij} \in Z_2.\qquad(21)$$

Let

$$C_0 = \{ j: a_{1j} = 0, 1 \leqslant j \leqslant p(2^p - 1) \},$$
$$C_1 = \{ j: a_{1j} = 1, 1 \leqslant j \leqslant p(2^p - 1) \}. \tag{22}$$

Claim. (C_0, C_1) is a 2-coloring of $\{ 1, \ldots, p(2^p - 1) \}$ with no mono-chromatic AP of length $p + 1$. Suppose that $\{ a, a + b, a + 2b, \ldots, a + pb \} \subseteq C_k$, $k = 0$ or 1. Set $\beta = \alpha^a$, $\gamma = \alpha^b$. Since $1 \leqslant a < a + pb \leqslant p(2^p - 1)$, $b < 2^p - 1$ so $\gamma \neq 1$. Then $\beta, \beta\gamma, \ldots, \beta\gamma^p$ have the same first coordinate as vectors.

CASE 1. $k = 0$. Then $\beta, \beta\gamma, \ldots, \beta\gamma^{p-1}$ are p vectors in a $(p - 1)$-dimensional space (since the first coordinate is 0), and hence they are dependent. Thus there exist $a_0, a_1, \ldots, a_{p-1} \in Z_2$, not all 0, such that

$$\sum_{i=0}^{p-1} a_i(\beta\gamma^i) = 0,$$

and hence

$$\sum_{i=0}^{p-1} a_i\gamma^i = 0.$$

But $\gamma \in GF(2^p)$, $\gamma \neq 0, 1$, so γ has degree p over $GF(2)$, a contradiction.

CASE 2. Assume that $\beta, \beta\gamma, \ldots, \beta\gamma^p$ have first coordinate 1. Now $\beta(\gamma - 1), \beta(\gamma^2 - 1), \ldots, \beta(\gamma^p - 1)$ lie in $(p - 1)$-dimensional space so

$$\sum_{i=0}^{p} a_i[\beta(\gamma^i - 1)] = 0,$$

where $a_i \in Z_2$, and some $a_i \neq 0$. Dividing by $\beta(\gamma - 1)$, we again find γ satisfying a polynomial of degree at most $p - 1$, a contradiction.

Theorem 7. $W(k) > [(2^k/2ek)(1 + o(1))].$

Proof. Randomly 2-color $[n]$, each i being colored red with probability $\frac{1}{2}$. For each AP S of k-terms let A_S be the event "S is monochromatic." Define a dependency graph, joining S and T iff $S \cap T \neq \emptyset$. If $n < (2^k/2e)(1 - \varepsilon)$ then, when Corollary 4 with $d = nk$ is applied, the event $\bigwedge_S \bar{A}_S$ has nonzero probability so that there is a 2-coloring of $[n]$ without monochromatic APs of size k.

The straightforward Existence argument in this instance would yield the much weaker result $W(k) > 2^{(k/2)(1+o(1))}$.

Asymptotic evaluation of $W(k, t)$ for fixed k is closed connected to the corresponding Turán problems. Define

$$\nu_k(n) = \max|S|: S \subset [n], \quad S \text{ does not contain an AP of } k\text{-terms.}$$

We shall consider only the case $k = 3$ and set $\nu(n) = \nu_3(n)$ for convenience.

Theorem 8. $ne^{-c\sqrt{\log n}} < \nu(n) < cn/\log\log n.$

The upper bound is due to K. Roth. The proof requires a relatively small modification of his proof given in Section 2.5 and is not presented here. The best current upper bound $\nu(n) < cn[(\log\log n)^5/\log n]$ is due to T. Sanders. We give the lower bound due to F. A. Behrend. For $d \geqslant 1$ we may write any a, $1 \leqslant a \leqslant n$, to the base $(2d + 1)$:

$$a = a_0 + a_1(2d + 1) + \cdots + a_k(2d + 1)^k, \qquad 0 \leqslant a_i \leqslant 2d.$$

Set

$$N(\mathbf{a}) = \left[\sum_{i=0}^{k} a_i^2\right]^{1/2}, \qquad \text{where } \mathbf{a} = (a_0, \ldots, a_k).$$

For $s \geqslant 1$ set

$$A = A_{n,d,s} = \{a: 1 \leqslant a \leqslant n, 0 \leqslant a_i \leqslant d \quad \text{for all } i, N(\mathbf{a})^2 = s\}. \quad (23)$$

For all n, d, s the set A contains no three-term arithmetic progression, for suppose that

$$a = \sum a_i(2d + 1)^i,$$

$$b = \sum b_i(2d + 1)^i,$$

$$c = \sum c_i(2d + 1)^i,$$

where all are in A and $a + b = 2c$. Since all $a_i, b_i, c_i \leqslant d$, there is no carrying in $a + b = 2c$ so $a_i + b_i = 2c_i$ for $0 \leqslant i \leqslant k$. Then

$$N(\mathbf{a}) = N(\mathbf{b}) = N[\tfrac{1}{2}(\mathbf{a} + \mathbf{b})],$$

which is possible only if $\mathbf{a}$ and $\mathbf{b}$ are proportional and, since $N(\mathbf{a}) = N(\mathbf{b})$, identical, that is, $\mathbf{a} = \mathbf{b} = \mathbf{c}$. The proof now becomes nonconstructive. For a given d

$$k \sim \frac{\log n}{\log(2d + 1)} ,$$

and there are at most d^2k possible values for s. The union of the $A_{n,d,s}$ over all s contains all sums $\Sigma\, a_i(2d + 1)^i \leq n$, $0 \leq a_i \leq d$. This is approximately $n2^{-k}$ elements. Consequently, for some s

$$\nu(n) \geq |A_{n,d,s}| \geq \frac{n}{d^2k2^k} . \tag{24}$$

Selecting d so that $k \sim \sqrt{\log n}$ maximizes the inequality, completing the proof.

Bounds of $W(3, t)$ based on Theorem 8 are given by Theorem 13.

4.4 THE SYMMETRIC HYPERGRAPH THEOREM

In Section 1.4 we compared Ramsey theorems and Density theorems. We noted that Density theorems of the appropriate form implied their corresponding Ramsey theorems. We now extend these results and show how results on density functions yield both upper and lower bounds on the corresponding Ramsey functions under appropriate circumstances.

Let $(S, \mathcal{Q})$ be a hypergraph. This means only that $\mathcal{Q}$ is a family of subsets of S. Assume that S is finite, set $m = |S|$, and assume that $\mathcal{Q}$ does not contain the null set or singleton sets. Call $T \subseteq S$ *free* if it contains no subset $A \in \mathcal{Q}$. Set:

$$\begin{aligned}
\nu &= \nu(S) = \text{the maximal } |T|, \quad T \subset S, T \text{ free}, \\
\chi &= \chi(S) = \text{the minimal } c \text{ so that one may partition} \\
S &= T_1 + \cdots T_c, \quad \text{all } T_i \text{ free.}
\end{aligned}$$

Here $\chi(S)$ is the usual definition of the chromatic number of a hypergraph. If S is χ-colored some color is used at least m/χ times. Hence we have the following theorem.

Theorem 9. $m/\nu \leq \chi$.

We call $(S, \mathcal{Q})$ symmetric if the automorphism group G of S is transitive. (A permutation σ of S is called an automorphism if $A \in \mathcal{Q}$ implies $\sigma A \in \mathcal{Q}$. The group G is transitive if, for all $s, s' \in S$, there exists $\sigma \in G$ so that $\sigma s = s'$.) For symmetric hypergraphs the following theorem allows us to use ν to get upper bounds on χ.

Theorem 10 (Symmetric Hypergraph Theorem).   If $(S, \mathcal{Q})$ is a symmetric hypergraph with m, ν, χ, G as defined above, then

$$m\left(1 - \frac{\nu}{m}\right)^{\chi - 1} \geq 1 ,$$

that is,

$$\chi \leq 1 + \frac{\log m}{-\log(1 - \nu/m)} . \tag{25}$$

We note that, for $\nu/m \ll 1$, we may use the approximation

$$\chi \leq \frac{m}{\nu} \log m . \tag{26}$$

Lemma.   Let U and T be arbitrary subsets of a symmetric hypergraph S with automorphism group G. There exists $\sigma \in G$ so that

$$|\sigma T \cap U| \geq \frac{|T||U|}{m} .$$

Proof.   We double-count triples (σ, t, u), $\sigma \in G$, $t \in T$, $u \in U$, so that $\sigma t = u$. We fix t, u. By the transitivity of G, $\sigma t = u$ for precisely $|G|/m$ automorphisms $\sigma \in G$. This gives precisely $|T||U||G|/m$ triples. For some fixed σ, at least $|T||U|/m$ pairs (t, u) satisfy $\sigma t = u$ and each pair has a distinct $u \in \sigma T \cap U$. (This proof does not construct an appropriate σ but only establishes its existence.)

Proof of Theorem 10.   Fix $T \subset S$, $|T| = \nu$, T free. Let r be that integer satisfying

$$m\left(1 - \frac{\nu}{m}\right)^{r} < 1$$

and

$$m\left(1 - \frac{\nu}{m}\right)^{r-1} \geq 1 . \tag{27}$$

We define a sequence $\sigma_1, \sigma_2, \ldots$ inductively. Having defined $\sigma_1, \ldots, \sigma_i$, we set

$$U_i = \{s \in S : s \neq \sigma_j t \quad \text{for } j \leq i, t \in T\} .$$

We define $U_0 = S$. By Lemma 11 we find σ_{i+1} so that

$$|\sigma_{i+1} T \cap U_i| \geqslant \frac{\nu}{m} |U_i| .$$

Therefore

$$|U_{i+1}| < \left(1 - \frac{\nu}{m}\right)|U_i|$$

for all i so that $U_r = \emptyset$. This implies that

$$S = \sigma_1 T \cup \cdots \cup \sigma_r T , \tag{28}$$

and, since each $\sigma \in G$ is an automorphism, each $\sigma_i T$ is free. The $\sigma_i T$ are not necessarily disjoint, but we may set $T_i = \sigma_i T - \cup_{j<i} \sigma_j T$ to get an r-coloration of S. Thus $\chi \leqslant r$, and Theorem 10 follows from the definition of r.

In asymptotic calculations we often have a sequence of hypergraphs $(S_m, \mathcal{D}_m)$, $|S_m| = m$ (not necessarily defined for all m), and functions $\nu(m)$, $\chi(m)$. Set

$$R_{\mathcal{D}}(t) = \text{minimal } m' \quad \text{so that, for } m \geqslant m', \chi(m) > t .$$

Define $f(m) = m/\nu(m)$ and $g(m) = f(m) \log m$. Assume that $f(m)$ tends to infinity. Then, essentially,

$$g^{-1}(t) \leqslant R_{\mathcal{D}}(t) \leqslant f^{-1}(t) . \tag{29}$$

The precise statement involves merely an unraveling of the definitions. The inequality (29) is correct within a $[1 + o(1)]$ factor if the sequence is reasonably smooth. Even when $f(m)$ is bounded a careful examination of the Symmetric Hypergraph theorem gives a lower bound for $R_{\mathcal{D}}(t)$, though there may be no upper bound since the corresponding Ramsey theorem may fail to hold.

In Graph Ramsey theory (see Section 5.7) the results obtained above are often useful. Let G be a finite graph. Set $T_G(n)$ equal to the maximal number of edges that a graph on n points may have and not contain a copy of G. Let $t_G(n) = T_G(n) \Big/ \binom{n}{2}$ for convenience. Let $\chi_G(n)$ equal the minimal number of colors required to edge-color K_n without forming a monochromatic G. We form a hypergraph $(S, \mathcal{D})$. Let $S = [n]^2$, that is, the edges of K_n are the vertices of S. A set $X \subseteq S$ is a hyperedge if X is the set of edges of a copy of G. $(S, \mathcal{D})$ is a symmetric hypergraph, as the full symmetric group on $[n]$ acts transitively on S. A direct application of the Symmetric Hypergraph theorem yields the following corollary.

Corollary 11. If $\lim t_G(n) = 0$ then

$$\frac{1}{t_G(n)} \leqslant \chi_G(n) < \frac{(1 + o(1)) \ln\binom{n}{2}}{t_G(n)}.$$

For example, when G is the 4-cycle it is known that $T_G(n) \sim cn^{3/2}$ so

$$cn^{1/2} < \chi_G(n) < cn^{1/2} \log n$$

and

$$\frac{ct^2}{(\log t)^2} < R_G(t) < ct^2,$$

where $R_G(t)$ is the minimal n so that if K_n is t-colored there always exists a monochromatic G.

Applying the Symmetric Hypergraph theorem to the function $W(3, t)$ of Section 4.3 requires some further preparation. We define a hypergraph $(S_n, \mathscr{D}_n)$ with $S_n = [n]$ and $A \in \mathscr{D}_n$ iff A is a three-term AP in $[n]$. We embed S_n into a symmetric hypergraph S'_n. The vertex set of S'_n is Z_{2n-1}. A set A is a hyperedge of S'_n iff A is a three-term AP in Z_{2n-1} and is contained in a block of n consecutive terms. For example, if $n = 50$, $\{3, 5, 7\}$ and $\{98, 1, 3\}$ are hyperedges but $\{0, 40, 80\}$ is not. Now the maps $\sigma_i: x \rightarrow x + i$ defined in Z_{2n-1} are automorphisms so that S'_n is a symmetric hypergraph.

From the original problem, let $\nu(n)$ be the maximal cardinality of a three-term progression-free subset of $[n]$, and $\chi(n)$ be the minimal number of colors required to color $[n]$ so that there is no monochromatic three-term AP. From Theorem 9,

$$\chi(n) \geqslant \frac{n}{\nu(n)} \tag{30}$$

immediately. Let $\nu'(n)$ and $\chi'(n)$ be the ν and χ values for the Z_{2n-1} hypergraph. First note that

$$\nu(n) \leqslant \nu'(n) \leqslant 2\nu(n). \tag{31}$$

The first inequality is immediate since S_n is a subhypergraph of S'_n. For the second, note that if T is free in S'_n then $T \cap \{1, \ldots, n\}$ and $T \cap \{n + 1, \ldots, 2n - 1\}$ have no three-term AP in Z so $|T| \leqslant \nu(n) + \nu(n - 1) \leqslant 2\nu(n)$.

Theorem 12. $\chi(n) < 2(n \log n)/\nu(n)(1 + o(1))$.

Proof. $\chi(n) \leq \chi'(n)$ as S_n is a subhypergraph of S'_n. By the Symmetric Hypergraph theorem

$$\chi'(n) < \frac{2n - 1}{\nu'(n)} \log(2n - 1)(1 + o(1))$$

$$< \frac{2n \log n}{\nu(n)} (1 + o(1)) \,.$$

Applying the bounds of Theorem 8 and (29), we obtain the following theorem.

Theorem 13. $t^{c \ln t} < W(3, t) < 2^{c't}$.

4.5 SCHUR AND RADO NUMBERS

Let $f(t)$ denote the maximal n so that it is possible to t-color $[n]$ with no monochromatic solution to the equation $x + y = z$. The finiteness of $f(t)$ is guaranteed by Schur's theorem. An examination of Proof 1 of Schur's theorem yields

$$f(t) \leq R(3, \ldots, 3) - 2 \,, \tag{32}$$

where R is the Ramsey function and there are t 3's. Schur notes that a t-coloring

$$[n] = C_1 + \cdots + C_t$$

without a monochromatic $x + y = z$ induces a similar coloring

$$[3n + 1] = C'_1 + \cdots + C'_t + C'_{t+1}$$

by setting

$$C'_i = C_i \cup (C_i + (2n + 1)) \,, \qquad 1 \leq i \leq t \,,$$
$$C'_{t+1} = \{n + 1, n + 2, \ldots, 2n + 1\} \,. \tag{33}$$

Since $f(1) = 1$ the construction outlined above gives $f(t) \geq (3^t - 1)/2$. It is not known whether $f(t)^{1/t}$ is bounded. The known exact values for f are $f(1) = 1$, $f(2) = 4$, $f(3) = 13$, and $f(4) = 44$, the last requiring a computer.

The evaluation of $f(5)$ appears to be a difficult computational problem.

For any integral m by n matrix A, let $f_A(t)$ denote the analogous function defined for the system $Ax = 0$ and let $f'_A(t)$ denote this function when a distinct solution (i.e., with $x_1, \ldots, x_n$ all different) is required. Rado's theorem (Chapter 3, Theorem 4, Corollary $8^{\frac{1}{2}}$) gives conditions for f_A and f'_A to be defined for all t. Few general results on the growth rate of these functions are known. If A is not regular no general means is known to determine the minimal t for which there exists a t-coloring of N without a monochromatic solution.

Let $v_A(n)(v'_A(n))$ be the maximal $|T|$, $T \subset [n]$, T not containing a solution (distinct solution) to $Ax = 0$.

Theorem 14. $v_A(n)/n \to 0$ iff $A1 = 0$. Furthermore, if $A1 = 0$ and there exists a distinct solution $(\lambda_1, \ldots, \lambda_m)$, then $v'_A(n) \to 0$.

Proof. If $A1 \neq 0$ then, for some $m \in N$, $A1 \not\equiv 0$ (modulo m). For any n,

$$T = \{i: 1 \leqslant i \leqslant n, i \equiv 1 \ (\text{modulo } m)\}$$

contains no solution so $v_A(n)/n \geqslant m^{-1}$. If $A1 = 0$ then $v_A(n) = 0$ since, for any a, $x_i = a$ gives a solution. Furthermore, if $(\lambda_1, \ldots, \lambda_m)$ is a distinct solution, set $k = 1 + \max|\lambda_i - \lambda_j|$. If $T \subseteq [n]$ contains an AP of length k it contains a distinct solution $x_i = a + \lambda_i d$. Now Szemerédi's theorem (Section 6.1) implies that $v_A(n) \to 0$.

In the special case $A = (1, 1, -1)$, the equation $x + y = z$, one can show that $v_A(n) = [(n + 1)/2]$.

4.6 PROPERTY B

Ramsey theory may be examined as the study of the chromatic number of certain hypergraphs. Valuable information may be gleaned from some general results on the chromatic number of hypergraphs. A hypergraph $\mathcal{A}$ is called n-uniform if all $A \in \mathcal{A}$ have $|A| = n$. Define $m(n)$ as the minimal cardinality of an n-uniform hypergraph with chromatic number >2. A hypergraph is said to have Property B if its chromatic number is $\leqslant 2$. It is known that $m(2) = 3$, $m(3) = 7$, the minimal hypergraphs being as follows:

$$\mathcal{A}_2 = \{\{1, 2\}, \{1, 3\}, \{2, 3\}\},$$

$$\mathcal{A}_3 = \{\{i, i + 1, i + 3\}, i \in Z_7, \text{ addition in } Z_7\}.$$

Evaluation of $m(4)$ also appears to be a difficult computational problem. Asymptotic lower bounds on $m(n)$ are given by an Existence argument.

Theorem 15. $m(n) \geq 2^{n-1}$.

Proof. Let $\mathcal{F} = \{S_1, \ldots, S_m\}$ be an n-family. Consider a random 2-coloring of $\cup \mathcal{F}$, each x independently colored red or blue with probability $\frac{1}{2}$. Since $|S_i| = n$,

$$\text{Prob}[S_i \text{ is monochromatic}] = 2^{1-n},$$
$$\text{Prob}[\text{some } S_i \text{ is monochromatic}) \leq m2^{1-n}.$$

For $m < 2^{n-1}$ this probability is less than unity; hence it must be possible to 2-color $\mathcal{F}$ so that no S_i is monochromatic.

The best asymptotic bounds currently known are

$$c2^n n^{1/3} \leq m(n) \leq c'2^n n^2 \tag{34}$$

Many Ramsey function bounds can be derived from Theorem 15. We examine $R(k)$ as an example. Given k and n, let $S = [n]^2$ and

$$\mathcal{F} = \{[T]^2 : |T| = k\}$$

the cliques of size k. Then $\mathcal{F}$ is a $\binom{k}{2}$-family, $|\mathcal{F}| = \binom{n}{k}$ so, if

$$m\left(\binom{k}{2}\right) > \binom{n}{k}, \tag{35}$$

the family $\mathcal{F}$ may be properly 2-colored, that is, $R(k) > n$.

The Lovász Local lemma has an important implication for Property B.

Theorem 16. Let $\mathcal{F}$ be an n-family. Suppose that every $S \in \mathcal{F}$ intersects at most d sets $T \in \mathcal{F}$. If

$$d + 1 < \frac{2^{n-1}}{e}$$

then $\mathcal{F}$ may be 2-colored.

Proof. Let $\mathcal{F}$ be 2-colored randomly as in the proof of Theorem 15. For $S \in \mathcal{F}$ let A_S be the event "S is monochromatic." Define a dependency graph joining A_S and A_T iff $S \cap T \neq \emptyset$. Apply Corollary 4 with $p = 2^{1-n}$. Then $P(\wedge \bar{A}_S) > 0$, and therefore $\mathcal{F}$ may be 2-colored.

Theorem 7 is an immediate corollary of the result obtain above.

4.7 HIGHER RAMSEY NUMBERS

We recall that $R_k(l)$ is the minimal n such that if $[n]^k = C_1 + C_2$ there exist i and S with $|S| = l$ and $[S]^k \subseteq C_i$. No exact values of $R_k(l)$ are known for $k \geq 3$. In this section we find asymptotic estimates of $R_k(l)$ for k fixed. We modify slightly the function TOWER of Section 2.7.

DEFINITION. The "tower functions" $t_i(x)$ are defined inductively by

$$t_1(x) = x ,$$
$$t_{i+1}(x) = 2^{t_i(x)} \tag{36}$$

so that, for example, $t_3(x) = 2^{2^x}$. For $k = 2$ we have shown that

$$(\sqrt{2} + o(1))^l < R_2(l) \leq (4 + o(1))^l .$$

For k fixed, the proof of Ramsey's theorem gives

$$\log_2 R_k(l) \leq R_{k-1}(l)^{k-1} \tag{37}$$

for l sufficiently large. (Actually this is a gross overestimate, but it sufficies for our purposes.) By induction

$$R_k(l) \leq t_k(c_k l) \tag{38}$$

for all $k \geq 2$. The lower bound requires the following lemma, which transforms a coloring of $[n]^k$ into a coloring of $[2^n]^{k+1}$.

Lemma 17 (Stepping-Up Lemma). If $n \nrightarrow (l)^k$ and $k \geq 3$ then

$$2^n \nrightarrow (2l + k - 4)^{k+1} .$$

Proof. Fix a 2-coloring $[n]^k = C_1 + C_2$ with no monochromatic l-element $S \subseteq [n]$. Set

$$T = \{(\gamma_1, \ldots, \gamma_n): \gamma_i = 0 \text{ or } 1\}.$$

For $\varepsilon = (\gamma_1, \ldots, \gamma_n)$, $\varepsilon' = (\gamma'_1, \ldots, \gamma'_n)$, $\varepsilon \neq \varepsilon'$, we define

$$\delta(\varepsilon, \varepsilon') = \max\{i: \gamma_i \neq \gamma'_i\}$$

and order T by [setting $i = \delta(\varepsilon, \varepsilon')$]:

$$\varepsilon < \varepsilon' \qquad \text{if } \gamma_i = 0, \gamma'_i = 1,$$
$$\varepsilon' < \varepsilon \qquad \text{if } \gamma_i = 1, \gamma'_i = 0.$$

The bijection between T and $[0, 2^n - 1]$ given by associating $(\gamma_1, \ldots, \gamma_n)$ with $\sum_{i=1}^n \gamma_i 2^{i-1}$ associates the above "$<$" on T with the "usual $<$." Note that:

(i) if $\varepsilon_1 < \varepsilon_2 < \varepsilon_3$ then $\delta(\varepsilon_1, \varepsilon_2) \neq \delta(\varepsilon_2, \varepsilon_3)$,
(ii) if $\varepsilon_1 < \varepsilon_2 < \cdots \varepsilon_n$ then $\delta(\varepsilon_i, \varepsilon_n) = \max_{1 \leq i \leq n-1} \delta(\varepsilon_i, \varepsilon_{i+1})$.

Now define a decomposition $[T]^{k+1} = I_1 + I_2$. Let $E = \{\varepsilon_1, \ldots, \varepsilon_{k+1}\}_< \in [T]^{k+1}$. Set $\delta_i = \delta(\varepsilon_i, \varepsilon_{i+1})$, $1 \leq i \leq k$. If the δ_i are monotonic (i.e., $\delta_1 < \delta_2 < \cdots < \delta_k$ or $\delta_1 > \delta_2 > \cdots > \delta_k$) place $E \in I$, iff $\{\delta_1, \ldots, \delta_k\} \in C_i$, that is, color the ε's by the δ's. If $\delta_1 < \delta_2 > \delta_3$ place $E \in I_1$. If $\delta_1 > \delta_2 < \delta_3$ place $E \in I_2$. The remaining E may be placed arbitrarily.

Let $S = \{\varepsilon_1, \varepsilon_2, \ldots, \varepsilon_{2l+k-4}\}_<$ be arbitrary. We assume that $[S]^k \subseteq I_1$, and derive a contradiction. Set $\delta_i = \delta(\varepsilon_i, \varepsilon_{i+1})$ for $1 \leq i < 2l + k - 4$.

CASE 1. There exists j so that the subsequence

$$\delta_j, \delta_{j+1}, \ldots, \delta_{j+l-1}$$

is monotonic. First assume that $\delta_j > \delta_{j+1} > \cdots > \delta_{j+l-1}$. Since this l-set cannot have all its k-subsets in the same class, there exist $j \leq i_1 < i_2 < \cdots < i_k \leq j + l - 1$ so that

$$\{\delta_{i_1}, \delta_{i_2}, \ldots, \delta_{i_k}\} \in C_2.$$

A contradiction is found by "stepping up" to the set

$$A = \{\varepsilon_{i_1}, \varepsilon_{i_2}, \ldots, \varepsilon_{i_k}, \varepsilon_{i_k+1}\}.$$

For $1 \leq t < k$,

$$\delta(\varepsilon_{i_1}, \varepsilon_{i_{t+1}}) = \max_{i_t \leqslant m < i_{t+1}} \delta_m$$

$$= \delta_{i_t} \quad \text{(by monotonicity)}$$

and

$$\delta(\varepsilon_{i_k}, \varepsilon_{i_k+1}) = \delta_{i_k}.$$

The δ_{i_t} are monotonic so A is colored "by the δ's" and $A \subset I_2$. If $\delta_1 < \delta_2 < \cdots < \delta_{j+l-1}$ the same argument holds with

$$A = \{ \varepsilon_{i_1}, \varepsilon_{i_1+1}, \varepsilon_{i_2+1}, \ldots, \varepsilon_{i_k+1} \}.$$

CASE 2 = NOT CASE 1.   For $2 \leqslant i \leqslant 2l - 3$ call i a local max if $\delta_{i-1} < \delta_i >$ δ_{i+1}, and a local min if $\delta_{i-1} > \delta_i < \delta_{i+1}$. There can be no local min i since then $\{ \varepsilon_{i-1}, \varepsilon_i, \varepsilon_{i+1}, \ldots, \varepsilon_{i+k-1} \} \in I_2$. Between any two local max's there must be a local min (a result well known to teachers of elementary calculus), and thus there is at most one local max i. Either $i \leqslant l - 1$ or $i \geqslant l$ or there is no i, but all roads lead back to Case 1.

Hence $[S]^k \nsubseteq I_1$. Similarly, $[S]^k \nsubseteq I_2$, completing the proof of Lemma 17. We let the reader test his or her understanding by seeing why $k \geqslant 3$ was required.

The best known bound for $R_3(l)$ is

$$R_3(l) \geqslant 2^{cl^2}, \tag{39}$$

proved by a simple Existence argument. By the Stepping-Up lemma we have the following theorem.

Theorem 18. $R_k(l) \geqslant t_{k-1}(c_k' l^2)$ for $k \geqslant 4$.

Open Problem.   Is $R_3(l) \geqslant t_3(cl)$?

An affirmative answer would imply that R_k is of the order t_k for all $k \geqslant 3$. The situation is surprisingly different if we allow four colors.

Theorem 19.   If $n \nrightarrow (l)^2$ then $2^n \nrightarrow (l+1)^3_4$.

Proof.   Fix a 2-coloring $[n]^2 = C_1 + C_2$ with no monochromatic l-element $S \subseteq [n]$. Define T, $\delta(\varepsilon, \varepsilon')$, and "$<$" as in the proof of Lemma 17. Our 4-coloring $[T]^3 = I_1 + I_2 + I_3 + I_4$ is defined by

$$\{\varepsilon_1, \varepsilon_2, \varepsilon_3\} \in I_1 \quad \text{iff } \{\delta_1, \delta_2\} \in C_1 \quad \text{and } \delta_1 < \delta_2,$$
$$\{\varepsilon_1, \varepsilon_2, \varepsilon_3\} \in I_2 \quad \text{iff } \{\delta_1, \delta_2\} \in C_1 \quad \text{and } \delta_1 > \delta_2,$$
$$\{\varepsilon_1, \varepsilon_2, \varepsilon_3\} \in I_3 \quad \text{iff } \{\delta_1, \delta_2\} \in C_2 \quad \text{and } \delta_1 < \delta_2,$$
$$\{\varepsilon_1, \varepsilon_2, \varepsilon_3\} \in I_4 \quad \text{iff } \{\delta_1, \delta_2\} \in C_2 \quad \text{and } \delta_1 > \delta_2.$$

Let $S = \{\varepsilon_1, \ldots, \varepsilon_{l+1}\}_<$ be arbitrary. We assume that $[S]^3 \subseteq I_1$, and derive a contradiction. (The other three cases are similar and are omitted.) Let $\delta_i = \delta(\varepsilon_i, \varepsilon_{i+1})$ for $1 \le i \le 1$. For $i \le l-1$, $\{\varepsilon_i, \varepsilon_{i+1}, \varepsilon_{i+2}\} \in I_1$ so

$$\delta_i = \delta(\varepsilon_i, \varepsilon_{i+1}) < \delta(\varepsilon_{i+1}, \varepsilon_{i+2}) = \delta_{i+1}.$$

The δ's thus form a monotonically increasing sequence. For arbitrary $1 \le i < j \le l$, $\{\varepsilon_i, \varepsilon_{i+1}, \varepsilon_{j+1}\} \in I_1$, and hence

$$\{\delta(\varepsilon_i, \varepsilon_{i+1}), \delta(\varepsilon_{i+1}, \varepsilon_{j+1})\} = \{\delta_i, \delta_j\} \in C_1.$$

Now $\{\delta_1, \ldots, \delta_l\}$ would form a monochromatic set. The contradicts our hypothesis on the coloring of $[n]^2$.

REMARKS AND REFERENCES

§1. Gleason and Greenwood [1955] give the first values of $R(k, l)$. Of the plethora of papers on this topic we mention Graham and Rödl [1987] (a survey with numerous references to earlier work); Giraud [1973]; Burling and Reyner [1972]; Kalbfleisch [1967], [1971]; Kalbfleisch and Stanton [1968]; Walker [1971]; Grinstead and Roberts [1982] and Exoo [1989]. For $R_3(4)$ see Isbell [1969] and Giraud [1969]. For $R(3:4)$ see Folkman [1974].

§2. Erdös [1947] may be regarded as the seminal paper in the development of the Existence argument. Erdös and Spencer [1974] discuss this argument in detail. The Lovász Local lemma is given in Erdös and Lovász [1975] and developed in Spencer [1975], [1977]. Ajtai, Komlós, and Szemerédi [1980] give Theorem 5 (upper bound). Erdös [1961], simplified by Spencer [1977], gives the lower bound. Frankl [1977] gives a constructive lower bound for $R(k, k)$.

§3. Asymptotic values for the van der Waerden function are given by Berlekamp [1968], Erdös and Rado [1952], and Moser [1960]. Exact values can be found in Chvátal [1970]. Roth [1953] gives the upper bound to $v_3(n)$. Lower bounds to $v_3(n)$ are given by Salem and Spencer [1942], Behrend [1946], and Moser [1953].

§4. We do not believe that the Symmetric Hypergraph theorem, and its applications to Graph Ramsey theory, have been published explicitly, though they have been part of the "folk literature" for some time.

§5.   Analytic bounds on $v_A(n)$ for certain A are given by Roth [1954], [1967] and Choi [1971].

§6.   Basic results on Property B appear in a series of papers by Erdös [1963a], [1964a], [1969]. The improved lower bound can be found in Beck [1978].

§7.   Erdös and Rado [1952] give explicit upper bounds for $R_k(n)$. Lower bounds for $R_k(n)$ are due to Erdös and Hajnal.

5

Particulars

5.1 BIPARTITE RAMSEY THEOREMS

Let $K_{m,n}$ denote the complete bipartite m by n graph; that is, $K_{m,n}$ consists of $m + n$ vertices, partitioned into sets of size m and n, and the mn edges between them. We gvie an analogue to Ramsey's theorem for bipartite graphs.

Theorem 1. For all a and r there exists m so that if $K_{m,m}$ is r-colored there exists a monochromatic $K_{a,a}$.

Theorem 1 is not unexpected, and Theorem 5 gives a much stronger result. What is surprising is that Theorem 1 may be proved as a Density theorem.

Theorem 2. For all integers a and all $\varepsilon > 0$ there exists m so that if G is a subgraph of $K_{m,m}$ with at least εm^2 edges then G contains a $K_{a,a}$.

Zarankiewicz [1951] defined $k_{a,b}(m, n)$ as the minimal e so that if G is a subgraph of $K_{m,n}$ and contains at least e edges then G contains a $K_{a,b}$. Alternatively, it is the minimal e so that any m by n 0–1 matrix with at least e 1's contains an a by b submatrix of all 1's. For convenience, define $k_a(n) = k_{a,a}(n, n)$.

Theorem 3. If $n\binom{e/n}{a} > (a-1)\binom{n}{a}$ then $k_a(n) \leq e$.

Proof. By a second derivative calculation, $\binom{x}{a}$ is a concave function for any positive integer a. Fix n, a, e satisfying the inequality. Let T (top) and B (bottom) be disjoint sets of n vertices. Let G be a subgraph of the $K_{n,n}$ defined on A, B. Assume that G has at least e edges. For $i \in T$, set

$$D_i = \{j \in B: \{i, j\} \in G\},$$
$$d_i = |D_i|,$$

so that $\sum_{i \in T} d_i \geq e$. Set

$$U = \{(i, X): X \subset B, |X| = a, X \subset D_i\}.$$

For each $i \in T$ there are precisely $\binom{d_i}{a}$ X's such that $(i, X) \in U$.

Now we use a general result on concave functions. If $f(x)$ is concave and $\bar{x} = (x_1 + \cdots + x_n)/n$, then

$$\sum_{i=1}^{n} f(x_i) \geq n f(\bar{x}).$$

We apply this result with $f(x) = \binom{x}{a}$:

$$|U| = \sum_{i \in T} \binom{d_i}{a} \geq n \binom{e/n}{a}.$$

For $X \subseteq B$, $|X| = a$ we set

$$T_X = \{i \in T: (i, X) \in U\}.$$

Then $|U| = \sum |T_X|$ so that, for at least one of the $\binom{n}{a}$ summands X,

$$|T_X| \geq \frac{|U|}{\binom{n}{a}} \geq \frac{n \binom{e/n}{a}}{\binom{n}{a}} > a - 1.$$

Let $T_X^* \subseteq T_X$ with $|T_X^*| = a$. Then $T_X^* \cup X$ is the desired K_{aa}.

For fixed a, $k_a(n) = o(n^2)$ so that Theorem 2, and hence also Theorem 1, hold.

The behavior of Zarankiewicz's function has been studied by many authors, including T. Kovari, V. Sós, and P. Turán [1954]; Erdös and Rado [1956]; Guy [1968], [1969]; Guy and Znam [1969]; and Chvátal [1969]. Erdös and Spencer [1974] discuss the asymptotic evaluation of k. Erdös and Moon [1964] show that if $K_{m,m}$ is 2-colored the fraction of $K_{a,b}$ which is monochromatic is, at least asymptotically, 2^{1-ab} for fixed a, b and m, n approaching infinity.

The k-partite analogue of Ramsey's theorem may also be proved as a density result.

Theorem 4. For all k, a, ε there exists n so that, if $G \subset A_1 \times \cdots \times A_k$, $|A_i| = n$ and if $|G| \geqslant \varepsilon n^k$ then there exist $B_i \subseteq A_i$, $|B_i| \geqslant a$ so that $B_1 \times \cdots \times B_k \subseteq G$.

We do not give the proof (which is similar to that of Theorem 3). The proof is given by Erdös [1964b]; an account is given by Erdös and Spencer [1974].

The following more general result is proved by a simple Induced Coloring argument.

Theorem 5. For all $k > 0$, $s_1, \ldots, s_k > 0$, $a_1, \ldots, a_k > 0$, and all $r > 0$, there exist $n_1, \ldots, n_k > 0$ so that, if $|B_i| \geqslant n_i$, $1 \leqslant i \leqslant k$, and $[B_1]^{s_1} \times \cdots \times [B_k]^{s_k}$ is r-colored, then there exist $A_i \subset B_i$ $|A_i| = a_i$ so that $[A_1]^{s_1} \times \cdots \times [A_k]^{s_k}$ is monochromatic.

Proof. Use induction on k. The case $k = 1$ is covered by Ramsey's theorem. Fix all parameters, and let $n_1, \ldots, n_{k-1}$ be defined inductively to meet the conditions of Theorem 5 for $k, s_1, \ldots, s_{k-1}, a_1, \ldots, a_{k-1}, r$. Define n_k so that

$$n_k \to (a_k)_M^{s_k}, \qquad \text{where } M = r^T, \; T = \binom{n_1}{s_1} \cdots \binom{n_{k-1}}{s_{k-1}}.$$

Let $|B_i| = n_i$ (if $|B_i| > n_i$ restrict attention to a subset of cardinality n_i), and let χ be an r-coloring of $[B_i]^{s_1} \times \cdots \times [B_k]^{s_k}$. Define a coloring χ' on $[B_k]^{s_k}$ by

$$\chi'(U) = \chi'(U')$$

iff

$$\chi((C_1, \ldots, C_{k-1}, U)) = \chi((C_1, \ldots, C_{k-1}, U')) \qquad \text{for all } C_i \in [B_i]^{s_i}.$$

Since χ' is an M-coloring, there exists $A_k \subset B_k$, $|A_k| = a_k$ so that $[A_k]^{s_k}$ is monochromatic under χ'. Define χ'' for $C_i \in [B_i]^{s_i}$ by

$$\chi''((C_1, \ldots, C_{k-1})) = \chi((C_1, \ldots, C_{k-1}, U)) \qquad \text{for any } U \in [A_k]^{s_k}.$$

By induction there exist $A_1, \ldots, A_{k-1}$ so that $[A_1]^{s_1} \times \cdots \times [A_{k-1}]^{s_{k-1}}$ is monochromatic under χ'' and therefore $[A_1]^{s_1} \times \cdots \times [A_{k-1}]^{s_{k-1}} \times [A_k]^{s_k}$ is monochromatic under χ.

One should be careful about infinite analogues to Theorem 1. Define

$$\chi: N \times N \to \{\text{red, blue}\}$$

by

$$\chi(i, j) = \begin{cases} \text{red} & \text{if } i \leq j, \\ \text{blue} & \text{if } i > j. \end{cases}$$

Clearly, there are no infinite subsets A, B so that χ is monochromatic on $A \times B$. However, this gives essentially the only "counterexample," as the following result shows.

Theorem 6. Let χ be a finite coloring:

$$\chi: N \times N \to [r].$$

Then there exists an infinite set $A = \{a_i\}_< \subset N$ and colors c_L, c_G, c_E (not necessarily distinct) so that

$$\chi(a_i, a_j) = \begin{cases} c_L & \text{if } i < j, \\ c_G & \text{if } i > j, \\ c_E & \text{if } i = j. \end{cases}$$

Also there exist infinite sets $B = \{b_i\}_<$, $C = \{c_i\}_<$ and colors c_{LE}, c_G (not necessarily distinct) so that

$$\chi(b_i, c_j) = \begin{cases} c_{LE} & \text{if } i \leq j, \\ c_G & \text{if } i > j. \end{cases}$$

Proof. We define a coloring χ' of $[N]^2$ by

$$\chi'(\{i, j\}_<) = (\chi(i, j), \chi(j, i), \chi(i, i)).$$

As χ' is a finite coloring (with r^3 colors), there exists an infinite set $A = \{a_i\}_< \subset N$ so that $[A]^2$ is monochromatic under χ'. But then A satisfies the first part of Theorem 6, and setting $b_i = a_{2i-1}$ and $c_i = a_{2i}$, we find that the sets $B = \{b_i\}$, $C = \{c_i\}$ satisfy the second part.

5.2 INDUCED RAMSEY THEOREMS

A graph $G = (V(G), E(G))$ is an *induced* subgraph of $H = (V(H), E(H))$ if $V(G) \subset V(H)$ and $E(G) = \{\{i, j\} \in E(H), i, j \in G\}$. An induced subgraph G consists of all edges of H on a subset of $V(H)$. For convenience, if G is isomorphic to an induced subgraph of H we call G an induced subgraph of H.

Theorem 1 (Vertex-Induced Graph Theorem). For all $G, r > 0$ there exists H so that if the vertices of H are r-colored there exists an induced subgraph G with vertices monochromatic.

Theorem 2 (Edge-Induced Graph Theorem). For all $G, r > 0$ there exists H so that if the edges of H are r-colored there exists an induced subgraph G with edges monochromatic.

We present the proof of Nešetřil and Rödl [1978a]. Another proof, using quite different techniques, is given by Deuber [1975b].

These results are immediate if the word "induced" is removed. Let $|V(G)| = v$, and set $H = K_N$, where $N \to (v)_r$. An r-coloring of the edges of K_N yields a monochromatic K_v and $G \subset K_v$. For vertex coloring we set $H = K_M$, where $M = (v - 1)r + 1$.

We begin by defining a special class of graphs $H_{n,m}$ for all $m, n > 0$. Let $|D| = m, |R| = n$. The vertices of $H_{n,m}$ are the n^m functions $f: D \to R$. Two vertices f, g are adjacent if they have no common point, that is,

$$\{f, g\} \in E(H_{n,m}) \quad \text{if } f(x) \neq g(x) \qquad \text{for all } x \in D .$$

Clearly, the graph $H_{n,m}$ is independent of the specific choice of D and R. More generally, let $|D| = m$, and let R_d be defined for each $d \in D$ so that $|R_d| = n$. We may define the vertices of $H_{n,m}$ as those functions f with domain D so that $f(d) \in R_d$ for each $d \in D$, with adjacency as before. This gives the same graph $H_{n,m}$. Let $A_d \subseteq R_d, |A_d| = n'$ for $d \in D$. The set of functions f such that $f(d) \in A_d$ for all $d \in D$ generates a subgraph of $H_{n,m}$: This subgraph is called the restriction of $H_{n,m}$ to $\{A_d\}$. It is isomorphic to $H_{n',m}$.

It may be helpful to think of the elements of $H_{n,m}$ as ordered m-tuples $(x_1, \ldots, x_m), x_i \in R$. Two m-tuples are then adjacent if they have no coordinate in common.

Lemma 3. For all G there exist n, m so that G is an induced subgraph of $H_{n,m}$.

Proof. Let $n = |V(G)|$. Let D be the family of functions γ;

$$\gamma: V(G) \to [n]$$

such that if $\{i, j\} \in E(G)$ then $\gamma i \neq \gamma j$. The set D can be considered the family of n-colorations of G. Set $m = |D|$. Let $R = [n]$. Define a map

$$\Psi: G \to H_{n,m} , \qquad \text{setting } \Psi(v) = \hat{v} ,$$

where

$$\hat{v}: D \to R$$

is defined by

$$\hat{v}(\gamma) = \gamma(v) .$$

Cleary, Ψ is injective. If $\{v, w\} \in E(G)$ then $\gamma v \neq \gamma w$ for all $\gamma \in D$ so that $\hat{v}(\gamma) \neq \hat{w}(\gamma)$ for all $\gamma \in D$, and hence $\{\hat{v}, \hat{w}\} \in E(H_{m,n})$. Conversely, if $\{v, w\} \notin E(G)$ there exists $\gamma \in D$ such that $\gamma v = \gamma w$. For example, one can color v, w identically and all other points with distinct colors. Then $\hat{v}(\gamma) = \hat{w}(\gamma)$ so $\{\hat{v}, \hat{w}\} \notin E(H_{m,n})$. Thus G is an induced subgraph of $H_{n,m}$.

It now suffices to prove the Induced Graph theorems for $G = H_{n,m}$. The vertex case is immediate. By Section 5.1, Theorem 5, there exists N so that, if $H_{N,m}$ is vertex r-colored, there exists a monochromatic $H_{n,m}$. (Here we are applying Section 5.1, Theorem 5, with all $s_i = 1$ and thinking of the elements of $H_{N,m}$ as ordered m-tuples.) To review, for any G, r we find n, m so that G is an induced subgraph of $H_{n,m}$, and N so that an r-coloring of $H_{N,m}$ yields a monochromatic $H_{n,m}$. Then $H = H_{N,m}$ is the desired graph.

It is the proof of the Edge-Induced Graph theorem that is truly remarkable. No other result in Ramsey theory makes quite as much use of the techniques of the subject.

We will think of $H_{N,M+1}$ with $D = \{0\} \cup [M]$. For $f, g \in H_{N,M+1}$ (i.e., $f, g: \{0\} \cup [M] \to [N]$) we write $f < g$ if $f(0) < g(0)$. This is not a total ordering, but adjacent edges are comparable. We define the type of an edge of H by

$$t(\{f, g\}_<) = \{i \in [M]: f(i) > g(i)\} .$$

An edge coloring of $H_{N,M+1}$ is called canonical if the color of an edge depends only on its type, that is, if

$$t(\{f, g\}) = t(\{f', g'\}) \Rightarrow \chi(\{f, g\}) = \chi(\{f', g'\}) .$$

Lemma 4. For all n, M, r there exists N so that if $H_{N,M+1}$ is edge r-colored there exists $H_{n,M+1} \subseteq H_{N,M+1}$ canonically colored.

Proof. Let χ be an edge r-coloring of $H_{N,M+1}$ (N to be determined). Let $A_0, \ldots, A_M \in [N]^2$, $A_i = \{a_i, b_i\}_<$. $H_{N,M+1}$ restricted to $(A_0, \ldots, A_M)$ is isomorphic to $H_{2,M+1}$, which consists of 2^M disjoint edges, one of each type. We define an r^{2^M}-coloring χ', coloring $(A_0, \ldots, A_M)$ by the color of the 2^M edges under χ. Formally, for each $S \subset [M]$ we define (dependent on $\{A_i\}$) an edge $\{f_S, g_S\}_<$ of $H_{N,M+1}$ of type S by

$$f_S(0) = a_0, \qquad g_S(0) = b_0,$$

$$f_S(i) = \begin{cases} b_i & \text{if } i \in S, \\ a_i & \text{if } i \notin S, \end{cases} \qquad g_S(i) = \begin{cases} a_i & \text{if } i \in S, \\ b_i & \text{if } i \notin S. \end{cases}$$

Then we set

$$\chi'(A_0, \ldots, A_M) = \chi'(A'_0, \ldots, A'_M)$$

if, for all $S \subset [M]$, $\chi\{f_S, g_S\}) = \chi(\{f'_S, g'_S\})$.

Now (and, formally, this is the beginning of the proof), we define N so that under any r^{2^M}-coloring of $[N]^2 \times \cdots \times [N]^2$ ($M + 1$ factors) there exist $B_0, \ldots, B_M$, $|B_i| = n$ so that $[B_0]^2 \times \cdots \times [B_M]^2$ is monochromatic. The existence of N follows from Section 5.1, Theorem 5. Given any edge r-coloring χ of $H_{N,M+1}$, we define χ' as above and find a subgraph $H_{n,M+1}(= H_{N,M+1}$ restricted to $B_0, \ldots, B_M)$ on which χ' is monochromatic. This subgraph is colored canonically, for let $\{f, g\}_< \in E(H_{n,M+1})$ of type S. Then if we set $A_i = \{f(i), g(i)\}_<$, $\{f, g\}_< = \{f_S, g_S\}_<$ so that $\chi(\{f, g\})$ depends only on S.

Lemma 5. For all n, m, r there exists M so that if $H_{n,M+1}$ is edge r-colored canonically there exists a monochromatic $H_{n,m+1} \subseteq H_{n,M+1}$.

Proof. We select M (by the Extended Hales–Jewett theorem) so that if $2^{[M]}$ is r-colored there exist disjoint $B_0, B_1, \ldots, B_m \subset [M]$, nonempty except possibly for B_0, so that all

$$B_0 \cup \bigcup_{i \in I} B_i, \qquad I \subseteq [m],$$

are colored the same. A canonical r-coloring χ of $H_{n,M+1}$ induces an r-coloring χ' of $2^{[M]}$, coloring $S \subset [M]$ by the color of all edges of type S. Let $B_0, B_1, \ldots, B_m$ be as above, and set $B' = [M] - B_0 - B_1 - \cdots - B_m$.

Let H denote the set of functions $f \colon \{0\} \cup [M] \to [n]$ (i.e., elements of $H_{n,M+1}$) satisfying the following conditions:

(i) f is constant on B_i for each $i, 1 \le i \le m$;
(ii) f is constant on $\{0\} \cup B'$;
(iii) f is constant on B_0 with value $n + 1 - f(0)$.

We first claim that $H \cong H_{n,m+1}$. For each $(x_0, x_1, \ldots, x_m) \in H_{n,m+1}$ we associate the $f \in H$ given by

$$f(a) = \begin{cases} x_1 & \text{if } a \in B_i, 1 \le i \le m, \\ x_0 & \text{if } a \in \{0\} \cup B', \\ n + 1 - x_0 & \text{if } a \in B_0. \end{cases}$$

Moreover, the edges of H are all of a special type. Let $\{f, g\}_< \in E(H)$, and set

$$S = t(\{f, g\}_<) = \{\alpha \in [M] \colon f(a) > g(a)\}.$$

For all $a \in B'$,

$$f(a) = f(0) < g(0) = g(a)$$

so that $a \notin S$. For all $a \in B_0$,

$$f(a) = n + 1 - f(0) > n + 1 - g(0) = g(a)$$

so that $a \in S$. As f and g are constant on B_i, either $B_i \subseteq S$ or $B_i \cap S = \varnothing$ for $1 \le i \le m$. Hence all types are of the form $B_0 \cup B_{i_1} \cup \cdots \cup B_{i_t}$ (where possibly $t = 0$). Since χ' is constant on these sets, χ is monochromatic on H.

Our proof of the Edge-Induced Graph theorem is now complete, but an overview is certainly in order. We fix G and $r > 0$. First, we find n, m so that G is an induced subgraph of $H_{n,m+1}$. Second, we find M so that if $H_{n,M+1}$ is canonically edge r-colored there is a monochromatic $H_{n,m+1}$. Third, we find N so that if $H_{N,M+1}$ is edge r-colored there exists a canonically colored $H_{n,M+1}$. $H = H_{N,M+1}$ is the desired graph. An r edge coloring of $H = H_{N,M+1}$ yields a canonical $H_{n,M+1}$ that in turn yields a monochromatic $H_{n,m+1}$ within which, snug as the proverbial bug in a rug, lies our monochromatic G.

5.3 RESTRICTED RAMSEY RESULTS

Let G, H be finite graphs. We write

$$G \rightarrow (H)^1_r$$

if, given any r-coloring of the vertices of G, there exists a monochromatic induced subgraph H. We write

$$G \rightarrow (H)^2_r$$

if, given any r-coloring of the edges of G, there exists a monochromatic induced subgraph H. In Section 5.2 we showed that, for all $H, r, i = 1, 2$, there exists $G, G \rightarrow (H)^i_r$. Here, using different techniques, we strengthen that result.

The first result in Ramsey theory may be written as $K_6 \rightarrow (K_3)^2_2$. P. Erdös first asked whether there existed $G, G \rightarrow (K_3)^2_2$ with $\omega(G)$ small. [We define $\omega(G)$, the clique number of G, as the size of the maximum complete subgraph in G.] Small G's with that property have been found for $\omega(G) = 5$ (Graham [1968]) and $\omega(G) = 4$ (Irving [1973]). Folkman [1970] constructed a gigantic graph G with that property and $\omega(G) = 3$. More generally, for all n he constructed $G, \omega(G) = n, G \rightarrow (K_n)^2_2$. Surprisingly, Folkman's construction worked only for two colors, and the question of whether, for all r, n, there existed $G, \omega(G) = n, G \rightarrow (K_n)^2_r$ remained open for several years. The work of J. Nešetřil and V. Rödl [1976] answers this question affirmatively. This result constitutes the main body of this section. In fact, the argument we give below for graphs can be directly extended to the more general case of hypergraphs. At the end of this section we make some comments about this extension.

In Nešetřil's and Rödl's argument the role of bipartite graphs is central. Let G be a bipartite graph with vertices $V(G) = V_1 \cup V_2$, $V_1 \cap V_2 = \varnothing$, and edges $E(G)$ such that $(x, y) \in E(G)$ only if $x \in V_1$, $y \in V_2$. Then we define the "diagonal power" $G^{(m)}$ as follows: $V(G^{(m)}) = V_1^m \cup V_2^m$, and $E(G^{(m)}) = \{((x_1, \ldots, x_m), (y_1, \ldots, y_m)) | (x_i, y_i) \in E(G), 1 \leq i \leq m\}$.

Lemma. For any bipartite graph G and integer $r > 0$, there is a bipartite graph H such that $H \rightarrow (G)^2_r$.

Proof. Let $N = HJ(r, e)$, the Hales–Jewett number (see Section 2.2) for r-colors and an e-element set, where $e = |E(G)|$. Then $H = G^{(N)}$ is the desired graph. We see this easily: By definition of $G^{(N)}$, there is a one-to-one correspondence between edges of $G^{(N)}$ and N-tuples of edges of G. Let $E(G) = \{E_1, E_2, \ldots, E_e\}$. Thus an r-coloring of $E(H)$ yields an r-coloring of $E(G)^N$. By the Hales–Jewett theorem, there will be a monochromatic "line" in $E(G)^N$. This is a set of e N-tuples $(E_{1,i}, E_{2,i}, \ldots, E_{N,i})$, $1 \le i \le e$, where, for each j, either $E_{j,i} = E_{j,i'}$ for all i, i' (we call these j's the "constant" j's), or $E_{j,i} = E_i$ for all i. Consider the vertices obtained from these edges, $U = U_1 \cup U_2$, where

$$U_1 = \{(x_1, \ldots, x_N) | x_j \text{ is the vertex of } E_{j,1} \text{ in } V_1$$
$$\text{if } j \text{ is a "constant," and } x_{j'} + x_{j''} \text{ for all}$$
$$j', j'' \text{ not "constant"}\},$$

and U_2 is defined similarly. The subgraph $G^* \subseteq H$ induced by these vertices is isomorphic to G and has exactly the edges corresponding to the monochromatic line in $E(G)^N$. Thus G^* is the desired monochromatic subgraph.

Theorem 1. For any $r > 0$ and graph G, with $\omega(G) \le n$, there is a graph H with $\omega(G) \le n$ such that $H \rightarrow (G)_r^2$.

Proof. The proof is obtained by an iterated construction. Let $V(G) = \{v_1, \ldots, v_m\}$. Then we start with a graph P_0 consisting of R levels, $L_1, \ldots, L_R$, of vertices, where $R = R(m; r)$, the Ramsey number of r-coloring edges of complete graphs. We form P_0 from $\binom{R}{m}$ disjoint copies of G as follows: For each choice of an m-subset $i_1 < i_2 < \cdots < i_m$ from $\{1, 2, \ldots, R\}$ we introduce new vertices $u_{i_j} \in L_{i_j}$, $1 \le j \le m$, and edges (u_{i_j}, u_{i_k}) iff (v_j, v_k) is an edge of G. In other words, P_0 consists of disjoint copies of G, one spread across each subset of m of the levels of P_0. For example, if G is a 4-cycle (see Fig. 5.1), $m = 4$, $R = R(4, 4) = 17$, and P_0 consists of $\binom{17}{4}$ disjoint 4-cycles spread over 17 levels.

We now describe how to get P_1 from P_0. Let B_1 be the subgraph of P_0 generated by L_1 and L_2. B_1 is bipartite. Using the lemma, we let A_1 be a bipartite graph with $A_1 \rightarrow (B_1)_r^2$. For each copy of B_1 in A_1, we adjoin vertices and edges to complete it to form a copy of P_0, all such copies being disjoint except for possible overlaps in A_1. Then P_1 is the R-level graph consisting of the union of A_1 and all of these P_0's. The jth level of P_1 is the union of the jth level of the P_0's, for $j \ge 3$, and the first and second levels of P_1 are just the two parts of A_1 (see Fig. 5.2). If the edges of P_1 are r-colored there exists, on levels 1 and 2, a monochromatic B_1 that may be extended to a copy of P_0 with the property that all edges between levels 1 and 2 are the same color.

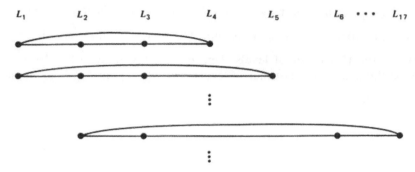

Figure 5.1 P_0 for $G = 4$-cycle.

We now iterate this construction, obtaining a sequence of R-level graphs $P_0, P_1, \ldots, P_{\binom{R}{2}}$. Let the $\binom{R}{2}$ pairs of levels be listed, say lexicographically, as $c_1, c_2, \ldots, c_{\binom{R}{2}}$. To get an R-level graph P_i from the graph P_{i-1}, we consider the ith pair c_i of levels of P_{i-1}, let B_i be the subgraph of P_{i-1} induced by these two levels, and let A_i be the bipartite graph such that $A_i \to (B_i)_r^2$. Then, for each copy of $B_i \subseteq A_i$, we adjoin disjoint completions to P_{i-1}, and the union of all of these is P_i. The last iteration of the process is $P_{\binom{R}{2}}$, and we claim that $H = P_{\binom{R}{2}}$ has the required properties.

We see this by sequentially applying the lemma. Let the edges of H be r-colored. If we consider just the last pair of levels, $\{m, m - 1\}$, in $P_{\binom{R}{2}}$, we get $A_{\binom{R}{2}}$. By the lemma and the choice of $A_{\binom{R}{2}}$, there is a monochromatic $B_{\binom{R}{2}} \subseteq A_{\binom{R}{2}}$. By construction of $P_{\binom{R}{2}}$, there is a copy

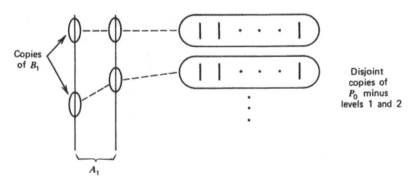

Figure 5.2 Construction of P_1 from P_0.

$P_{\binom{R}{2}-1} \subseteq P_{\binom{R}{2}}$ whose last pair of levels is precisely this $B_{\binom{R}{2}}$. Thus this (induced) subgraph of $P_{\binom{R}{2}}$ is isomorphic to $P_{\binom{R}{2}-1}$ and has all edges between the (last) pair of levels, $\{m, m-1\}$, the same color. Now we look at the next to the last pair of levels, $\{m, m-2\}$, in this $P_{\binom{R}{2}-1}$, and by the same argument get an induced copy of $P_{\binom{R}{2}-2}$ in it with all edges between these levels the same color. The edges of $P_{\binom{R}{2}-2} \subseteq P_{\binom{R}{2}-1}$ between the levels m and $m-1$ all have one color, and the edges between levels m and $m-2$ all have one color, possibly different from the color for levels $m, m-1$.

By repeating this argument $\binom{R}{2}$ times, we eventually obtain a copy of P_0 as an induced subgraph of H, such that the color of an edge depends only on the pair of levels the edge connects. This induces a coloring of the pairs of levels, each pair having the color of (all) the edges between them. By the choice of R, some subset of m levels has all pairs the same color. Finally, by construction of P_0, some copy of G is contained exactly among this set of m levels. Hence G is monochromatic. This completes the proof that $H \to (G)_r^2$. To complete the proof of Theorem 1 we need only observe that $\omega(H) = \omega(G)$, since at each step in the process of constructing H we never introduce cliques larger than those in G.

In fact, we observe more here. Let K be any graph for which there is no vertex cut-set inducing a subgraph of chromatic number smaller than 3. (A cut-set is a set of vertices whose removal leaves a graph with more than one connected component.) Call such a K 3-chromatic connected.

Theorem 2. Let K be 3-chromatic connected, and let G be a graph containing no K. Then there in an H containing no K such that $H \to (G)_r^2$.

The proof is the same as that for Theorem 1. We simply note that at each step in the construction we can never introduce a subgraph of type K, since all the copies of P_{i-1} overlap in a bipartite graph.

As mentioned earlier, Theorem 1 extends directly to hypergraphs. To prove it we need the hypergraph version of the lemma, which is simply the same result for a-uniform, a-partite hypergraphs rather than for bipartite ordinary graphs. The proof is essentially the same. Then the proof of the hypergraph result (Theorem 3 below) is exactly analogous to that of Theorem 1, where we have many copies of hypergraphs P_{i-1} from the $(i-1)$st step, disjoint except for possible overlap in a set of a levels at the ith step, P_i.

Theorem 3. Let G be an a-uniform hypergraph, and $r > 0$ and integer. Then there is an a-uniform hypergraph H such that $H \to (G)^a_r$ and $\omega(H) = \omega(G)$; that is, if the edges of the hypergraph H are r-colored, there is a monochromatic induced subhypergraph of type G. [The clique nubmer $\omega(H)$ is the maximum number m of points so that all $\binom{m}{a}$ a-subsets are edges.]

Just as there is an analogue to Ramsey's theorem for vector spaces over finite fields (Section 2.4, Corollary 10) so there is an analogue to the induced and restricted result of Theorem 3 for vector spaces. To state it we essentially just replace the terms "set" and "subset" by the terms "space" and "subspace," respectively. More formally, we can define an a-uniform space system H to be a family of a-dimensional subspaces, the "edges," of some finite dimensional vector space V over a finite field F. In fact, H can be thought of as a q^a-uniform hypergraph on the set of vectors of V, where edges are required to be subspaces. Then an induced subsystem G of H is simply an a-uniform space system G that is an induced subhypergraph of H. Now let $\omega(H)$ denote the clique number of H, by which we mean the largest dimension m of a subspace U of V so that all of the a-dimensional subspaces of U are edges of H. Frankl, Graham, and Rodl [1987] and Promel [1986] give the following induced and restricted vector space version of Ramsey's Theorem.

Theorem 4. Let G be an a-uniform space system and r a positive integer. Then there is an a-uniform space system H such that $H \to (G)^a_r$ and $\omega(H) = \omega(G)$. That is, if the edges of H are r-colored there is a monochromatic induced subsystem $G' \cong G$. (We say $G' \cong G$ if $\dim(U) = \dim(U')$, where U, U' are the vector spaces for G, G', respectively, and there is a bijective linear map from U to U' such that the image A' of any a-dimensional subspace A of U is an edge of G' if and only if A is an edge of G.)

5.4 EQUATIONS OVER ABELIAN GROUPS

The results of this section are from Deuber [1975a].

Let G be an Abelian group. Let A be an m by n matrix with integral coefficients. We say that A is *partition regular in G* if, for every finite coloration of $G - \{0\}$, there is a monochromatic solution $x_1, \ldots, x_n$ to the system $Ax = 0$.

By "solution" in this section we shall mean a monochromatic solution to $Ax = 0$.

Recall that A satisfies the *Columns condition* if one can order the column vector $c_1, \ldots, c_n$ and find $1 \leqslant k_1 < k_2 < \cdots < k_t = n$ such that, setting

$$A_i = \sum c_\gamma, \quad \text{summed over } k_{i-1} < \gamma \leqslant k_i,$$

we have

(i) $A_1 = 0$,

and

(ii) for $1 < i \leqslant t$, A_i may be expressed as a linear combination (over Q) of $c_1, \ldots, c_{k_{i-1}}$.

Then A is partition regular in Z if A satisfies the Columns condition. For p prime, we say that A satisfies the p-Columns condition if (i) and (ii) hold when their summations and linear combinations are taken modulo p. Let:

$$T(A) = \{ p: A \text{ satisfies the } p\text{-Columns condition} \},$$
$$T'(A) = \{ p: x_1 = \cdots = x_n = 1 \text{ is a solution to } Ax = 0 \text{ (modulo } p) \}.$$
$$\text{Note that } T'(A) \subseteq T(A) \text{ by setting } t = 1, k_1 = n.$$

One can show, similarly to the proof of Chapter 3, Lemma 6, that if A does not satisfy the Columns condition, $T(A)$ is finite, and if A satisfies the Columns condition, $T(A)$ is cofinite. The equation $3x - 3y - z = 0$ [i.e., the matrix $A = (3, -3, -1)$] provides an example of a matrix satisfying the Columns condition but not satisfying the 3-columns condition.

Let Z_p^ω be the countable infinite-dimensional vector space over Z_p, considered as an Abelian group, that is, the elements are infinite sequences

$$(a_1, a_2, \ldots), a_i \in Z_p, \text{ all but finitely many } a_i = 0,$$

and addition is defined componentwise modulo p. Note (from elementary group theory) that $G \subseteq Z_p^\omega$ iff G contains an infinite number of elements of order p.

Theorem 1. A is partition regular in G iff one of the following holds:

(i) G contains an element α of order $p \in T'(A)$.

(ii) $G \supseteq Z_p^\omega$ for some $p \in T(A)$.

(iii) A satisfies the Columns condition and G contains either an element of infinite order or elements of arbitrarily high order.

Proof. We first show that any of conditions (i), (ii), or (iii) implies that A is partition regular in G.

(i) $x_1 = \cdots = x_n = \alpha$ is always a solution, regardless of the coloring.

(ii) In Z_p we find a parametric solution to the equation $A\mathbf{x} = \mathbf{0}$, given by

$$x_i = \sum_{\gamma=1}^{t} \lambda_{ij} v_j, \qquad 1 \le i \le n,$$

where each $(\lambda_{i1}, \ldots, \lambda_{it})$ has the first nonzero term $\lambda_{ij} = c$, a nonzero constant. (The method illustrated for Z in Section 2.3 suffices here.) For convenience, we assume that $c = 1$ by changing all λ_{ij} to λ_{ij}/c. Now we let G be finitely colored and restrict our attention to Z_p^ω, with a specified ordered basis. For

$$\mathbf{v} = (v_1, v_2, \ldots) \in Z_p^\omega - \{0\}$$

we set

$$\lambda(\mathbf{v}) = \text{first } v_i \text{ such that } v_i \ne 0,$$

$$n(\mathbf{v}) = \text{first } i \text{ such that } v_i \ne 0.$$

We define a finite coloring of the one-dimensional subspaces of Z_p^ω, coloring a subspace $\{\mathbf{v}, 2\mathbf{v}, \ldots, (p-1)\mathbf{v}\}$ by the color of the unique $s\mathbf{v}$ with $\lambda(s\mathbf{v}) = 1$. By the Vector Ramsey theorem (Section 2.4; we do not require the full strength of the theorem here) there exists $S \subseteq Z_p^\omega$, $S \cong Z_p^t$, monochromatic under the induced coloring. By standard linear algebra techniques we find a basis $\mathbf{w}_1, \ldots, \mathbf{w}_t$ for S so that

$$\lambda(\mathbf{w}_j) = 1, \qquad 1 < j \le t,$$

$$n(\mathbf{w}_1) < n(\mathbf{w}_2) < \cdots < n(\mathbf{w}_t).$$

Now

$$x_i = \sum_{j=1}^{t} \lambda_{ij} \mathbf{w}_j, \qquad 1 \le i \le n,$$

gives the desired monochromatic solution.

(iii) Let $G - \{0\}$ be k-colored. If N is k-colored then, by Rado's theorem, there will exist a monochromatic solution to $A\mathbf{x} = \mathbf{0}$. By

the Compactness principle, for some m and k-coloration of $[m]$
we also have a solution. Let $\alpha \in G$ have order at least $m + 1$
(possibly infinite). Then a solution is found within the set
$\{\alpha, 2\alpha, \ldots, m\alpha\}$.

We prove the converse (the necessity of the conditions) only in the
case where A does not satisfy the Columns condition and has exactly one
row. Let $A = (c_1, \ldots, c_n)$ so that $Ax = 0$ is the single equation

$$c_1 x_1 + \cdots + c_n x_n = 0 ,$$

where no subset of $\{c_i\}$ sums to zero.

Lemma 2. Let A be as above, and let $T(A)$ and $T'(A)$ be defined as
before. Then there exists u, dependent only on A, with the following
property: Let m be such that $(m, p) = 1$ for all $p \in T(A)$. Then $Z_m - \{0\}$
may be u-colored without forming a monochromatic solution to $Ax = 0$
(modulo m).

We first illustrate these ideas with the equation

$$x_1 + x_2 + x_3 + x_4 = 0 \quad [\text{i.e., } A = (1, 1, 1, 1)] .$$

here $T'(A) = \{2\}$, $T(A) = \{2, 3\}$. [Incidentally, note the parameteric
solution $x_1 = a$, $x_2 = x_3 = a + d$, $x_4 = d$ (modulo 3).] We can 4-color
$Z - \{0\}$, using the smod 5 coloring. This coloring may not apply to
$Z_m - \{0\}$ since solutions of the form $x_1 + x_2 + x_3 + x_4 = \lambda m$ might appear. Let m be large, $(m, 2) = (m, 3) = 1$. Write

$$Z_m = \left[-\frac{m-1}{2}, \frac{m-1}{2} \right] .$$

Split Z_m into five intervals

$$I_i = (a_i - R, a_i + R) ,$$

where $R \sim m/10$ and $|a_i| < m/2$. We now color each interval I_i without
forming a monochromatic solution to $x_1 + x_2 + x_3 + x_4 \equiv 0$ (modulo m).
Since $R < m/8$ there is at most one λ for which the equation $x_1 + x_2 +
x_3 + x_4 = \lambda m$ (addition in N) has a solution in I_i. Since $|x_j| \leqslant m/2$,
$|\lambda| \leqslant 2$. If $\lambda = 0$ we apply the smod 5 coloring to I_i. If $\lambda \neq 0$ we color $x \in I_i$
by the residue class of x (modulo 4). If a monochromatic solution
$x_1 \equiv x_2 \equiv x_3 \equiv x_4$ (modulo 4) existed in I_i then $4|x_1 + x_2 + x_3 + x_4 = \lambda m$.

This is impossible since $(4, m) = 1$ and $0 < |\lambda| < 4$. Finally, if no λ exists we 1-color I_i trivially.

Each I_i has required at most four colors. We distinguish the colors used in each interval. This gives a 20-coloring (or better) of $Z_n - \{0\}$ with the desired property.

Proof of Lemma 2. Let t be such that $Z - \{0\}$ may be t-colored. Set

$$C = |\Sigma\, c_i| \neq 0 \qquad \text{by assumption,}$$

$$D = \Sigma\, |c_i|, \ R = \left[\frac{m}{2D} \right].$$

Let $m > m'$, where m' is to be determined. We write $Z_m = ([-m/2], [m/2])$, and split Z_m into consecutive intervals of length $2R - 1$ (the last interval perhaps smaller), $I_i = (a_i - R, a_i + R)$, $1 \leq i \leq s$. We may do this with

$$s = \left\{ \frac{m}{2[m/2D] - 1} \right\}.$$

Now we select m' so that, for $m > m'$, $s \leq D + 1$. We color each $I_j - \{0\}$ separately. If $x_1, \ldots, x_n \in I_j$ we set

$$x_i = a_j + y_i, \qquad |y_i| < R$$

(calculations done in Z), and note that

$$\Sigma\, c_i x_i = (\Sigma\, c_i) a_j + \Sigma\, c_i y_i,$$

where

$$|\Sigma\, c_i y_i| < R(\Sigma\, |c_i|) \leq \frac{m}{2}.$$

In other words, $\Sigma\, c_i x_i$ lies within an interval of length m so there is at most one λ (dependent on j) for which $\Sigma\, c_i x_i = \lambda m$ is possible. Since $|a_j| < m/2$,

$$|\Sigma\, c_i x_i| \leq |\Sigma\, c_i||a_j| + |\Sigma\, c_i y_i| < \frac{m(C + 1)}{2}$$

so that $|\lambda| < (C + 1)/2 \leq C$.

If $0 < |\lambda| < C$ we color $x \in I_j$ according to its residue class modulo C. A monochromatic solution would imply that $C | \Sigma\, c_i x_i = \lambda_m$, which is impossible since $(C, m) = 1$. If $\lambda = 0$ we t-color $I_j - \{0\}$, using the coloring of $Z - \{0\}$. If no λ exists we 1-color I_j trivially.

Each interval has used at most $\max(C, t)$ colors so that $Z_m - \{0\}$ is colored with $s \max(C, t) \leqslant (D + 1) \max(C, t)$ colors. Finally, if $m \leqslant m'$ we can color each point distinctly, using at most $m' - 1$ colors. Thus all of $Z_m - \{0\}$ may be u-colored, where

$$u = \max(m' - 1, (D + 1) \max(C, t)) .$$

Lemma 3.   Let $H_i - \{0\}$ be s_i-colorable for $1 \leqslant i \leqslant k$. Then the product $H_1 \times \cdots \times H_k - \{0\}$ may be $(s_1 + \cdots + s_k)$-colored.

Proof.   Let χ_i denote the s_i-coloring of H_i with color sets assumed to be distinct. Define χ on $H_1 \times \cdots \times H_k - \{0\}$ by

$$\chi((h_1, \ldots, h_k)) = \begin{cases} \chi_1(h_1) & \text{if } h_1 \neq 0 , \\ \chi_2(h_2) & \text{if } h_1 = 0, h_2 \neq 0 , \\ \chi_i(h_i) & \text{if } h_1 = \cdots = h_{i-1} = 0, h_i \neq 0 . \end{cases}$$

Proof of Theorem 1 (Necessity, Limited Case).   Fix G and A. Let q be such that $Z - \{0\}$ may be colored by the smod q coloring. Let $T(A)$ and $T'(A)$ be as before. Let t be the number of $g \in G$ of order $\Pi\, p_i^{\alpha_i}$, where all $p_i \in T(A)$. Since $G \not\supseteq Z_p^\omega$ for all $p \in T(A)$, t is finite. Let u be as given in Lemma 2. We claim that $G - \{0\}$ may be β-colored where $\beta = t + (q - 1) + u$.

By compactness, it suffices to β-color any finite subset of $G - \{0\}$. We do somewhat more, β-coloring any finitely generated subgroup $H \subset G$. By the fundamental theorem of Abelian groups we can write

$$H = H_1 \times Z_m \times Z^v ,$$

where v is finite, $(m, p) = 1$ for all $p \in T(A)$, and all $h \in H_1 - \{0\}$ have order $\Pi_{p_i}^{\alpha_i}$, where all $p_i \in T(A)$. Now

$H_1 - \{0\}$ is t-colored since $|H_1 - \{0\}| \leqslant t$,

$Z_m - \{0\}$ is u-colored by Lemma 1, and

Z^v is $(q - 1)$-colored, coloring $(x_1, \ldots, x_v)$ by the smod q coloring of the first nonzero x_i.

Thus, by Lemma 3, $H - \{0\}$ is β-colored, completing the proof.

A reformulation of Theorem 1 makes it clear that there are only three reasons for A to be partition regular in G.

Theorem 1'. A is partition regular over G iff one of the following holds:

(i) For some $\alpha \in G - \{0\}$, A is partition regular over $\{\alpha\}$ (i.e., $x_1 = \cdots = x_n = \alpha$ is a solution).

(ii) For some prime p, A is partition regular in Z_p^ω and $G \supseteq Z_p^\omega$.

(iii) A is partition regular in Z, and G contains either Z or elements of arbitrarily high order.

5.5 CANONICAL RAMSEY THEOREMS

The results of this section are due to Erdös and Rado [1950].

In this section we consider colorations in which no restrictions (not even finiteness) is put on the number of colors used. When elements of a set are colored we have a simple result. Call a coloring χ of a set S canonical if χ is either

(i) monochromatic [i.e., $\chi(s) = \chi(t)$ for all $s, t \in S$],

or

(ii) distinct [i.e., $\chi(s) \neq \chi(t)$ for all $s, t \in S, s \neq t$].

Theorem 1. If an infinite set S is colored then some infinite subset T is canonically colored. For all k, if $|S| > (k-1)^2 + 1$ and S is colored there exists a subset $T \subseteq S$, $|T| = k$ that is canonically colored.

The situation when pairs are colored is more interesting. Let us restrict our attention to edge colorings. We write $\chi(i, j)$ for $\chi(\{i, j\}_<)$. We distinguish four special colorings of an ordered set S:

(i) distinct: $\chi(\{i, j\}) = \chi(\{k, l\})$ iff $\{i, j\} = \{k, l\}$,

(ii) min: $\chi(\{i, j\}) = \chi\{k, l\})$ iff $\min(i, j) = \min(k, l)$,

(iii) max: $\chi(\{i, j\}) = \chi(\{k, l\})$ iff $\max(i, j) = \max(k, l)$,

(iv) monochromatic: χ is constant.

A coloring χ is called *canonical* on S if it has one of these four properties. Note that if χ is canonical on S it is canonical (of the same type) on every subset $T \subset S$.

Theorem 2. For every coloration of $[N]^2$ there exists an infinite $T \subset N$ on which χ is canonical. For all k there exists n so that, for every coloration of $[n]^2$, there is a $T \subset [n], |T| = k$ on which χ is canonical.

Proof. Let χ be a coloring of $[N]^2$. To each $\{a_1, a_2, a_3, a_4\}_< \in [N]^4$ we associate an equivalence relation $\equiv$ on $[4]^2$ given by

$$\{i, j\} \equiv \{k, l\} \qquad \text{if } \chi(\{a_i, a_j\}) = \chi(\{a_k, a_l\}).$$

We define a coloring χ' on $[N]^4$ by defining $\chi'(\{a_1, a_2, a_3, a_4\}_<)$ to be the equivalence relation to which it corresponds. In other words, χ' is the set of equalities among the six edges of $\{a_1, a_2, a_3, a_4\}_<$ under χ. There are 203 possible values (colors) for χ', corresponding to the 203 partitions of a six-element set. (For example, one color class consists of all those $\{a_1, a_2, a_3, a_4\}_<$ on which the six colors are distinct. Another consists of all those for which $\chi(\{a_1, a_3\}) = \chi(\{a_2, a_4\})$ and the other four edges are colored with distinct colors.) By Ramsey's theorem there exist $N' \subset N$, N' infinite, on which χ' is constant. Let us assume that $N' = N$ for convenience, since only the cardinality and ordering of N' are important.

Now we reduce the possibilities from 203 to 4. Consider the possible equalities:

(*)
$$\chi(a_1, a_2) = \chi(a_1, a_3),$$
$$\chi(a_1, a_3) = \chi(a_1, a_4),$$
$$\chi(a_1, a_4) = \chi(a_1, a_2),$$
$$\chi(a_2, a_3) = \chi(a_2, a_4).$$

Each of these qualities holds either for all $\{a_1, a_2, a_3, a_4\}_<$ or for none of them. Plugging in the values (in order) $\{2, 4, 6, 7\}$, $\{2, 3, 4, 6\}$, $\{2, 4, 5, 6\}$, $\{1, 2, 4, 6\}$ for $\{a_1, a_2, a_3, a_4\}_<$, we find that ech of the equalities given above is equivalent to the statement $\chi(2, 4) = \chi(2, 6)$. Thus the equalities either all hold or all do not hold. Similarly the equations

(**)
$$\chi(a_1, a_4) = \chi(a_2, a_4),$$
$$\chi(a_2, a_4) = \chi(a_3, a_4),$$
$$\chi(a_3, a_4) = \chi(a_1, a_4),$$
$$\chi(a_1, a_3) = \chi(a_2, a_3)$$

either all hold or all do not hold.

If systems (*) and (**) both hold then all $\chi(a, b) = \chi(1, b) = \chi(1, 2)$ and χ is monochromatic. If system (*) holds then $\chi(i, j) = \chi(i, k)$ for all $i < j, k$. Suppose that χ is not the "min" coloring; then, for some $i \neq i'$ and j, k, $\chi(i, j) = \chi(i', k)$. Let $t = \max(j, k)$. By (*), $\chi(i, t) = \chi(i, j) = \chi(i', k) = \chi(i', t)$. But then system (**) will hold. Hence, if (*) holds but (**) does not, χ is a "min" coloring. Similarly, if (**) holds but (*) does not, χ is a "max" coloring.

Finally, assuming that neither (*) nor (**) hold, we need to show that χ is distinct. We illustrate this with only a single case. Assume that $\chi(1, 2) = \chi(3, 4)$. Then, for all $\{a_1, a_2, a_3, a_4\}_<$, $\chi(a_1, a_2) = \chi(a_3, a_4)$. Hence $\chi(3, 4) = \chi(1, 2) = \chi(3, 5)$, which is impossible since (*) does not hold.

The finite version uses essentially the same proof. For all k we select n so that

$$n \rightarrow (\max(k, 7))_{203} \, .$$

For any coloring χ of $[n]^2$ we find $T \subseteq [n]$, $|T| = \max(k, 7)$, on which χ' is constant and hence χ is canonical.

There is a simply stated generalization to r-tuples: A coloring χ of $[S]^r$, S ordered, is called canonical if, for some $V \subset \{1, \ldots, r\}$, $\chi(\{a_1, \ldots, a_r\}_<) = \chi(\{b_1, \ldots, b_r\}_<)$ iff $a_i = b_i$ for all $i \in V$.

Theorem 3. For every coloring of $[N]^r$ there exists an infinite $T \subset N$ on which the coloring is canonical. For all k and r, there exists n so that, for every coloring of $[n]^r$, there exist $T \subset [n]$, $|T| = k$, on which the coloring is canonical.

The proof, although following the main ideas of the case $r = 2$, is more complicated and will not be given here.

Let χ now be a coloring of the bipartite graph $A \times B$. We define four special colors.

(i) monochromatic: all $\chi(a, b)$ equal;

(ii) column: $\chi(a, b) = \chi(c, d)$ iff $a = c$ (i.e., only points in the same column are colored the same);

(iii) row: $\chi(a, b) = \chi(c, d)$ iff $b = d$ (i.e., only points in the same row are colored the same);

(iv) distinct: $\chi(a, b) = \chi(c, d)$ iff $a = c$ and $b = d$ (i.e., all points have distinct colors).

We say that χ is *canonical* on $A \times B$ if it is of one of these four types. Note that if χ is canonical on $A \times B$ and $A_1 \subset A, B_1 \subset B$ then χ is canonical of the same type on $A_1 \times B_1$.

Theorem 4. For any coloring χ of $N \times [2r^2 + 1]$ there exists $N_1 \subset N, N_1$ infinite, and $B' \subset B, |B'| = r + 1$, such that χ is canonical on $N_1 \times B'$.

Proof. For each $n \in N$ the nth column contains either $r + 1$ points colored the same or $2r+1$ points colored distinctly. As there are only a finite

$$\left(=\binom{2r^2+1}{r+1}+\binom{2r^2+1}{2r+1}\right)$$

number of choices for the row coordinates of these points, there exist $N_1 \subseteq N$, N_1 infinite, and $B \subseteq [2r^2+1]$ so that either

(i) $|B| = r+1$ and $\chi(n, b_1) = \chi(n, b_2)$ for all $n \in N_1$,

or

(ii) $|B| = 2r+1$ and $\chi(n, b_1) \neq \chi(n, b_2)$ for all $n \in N_1$, $b_1 \neq b_2 \in B$.

In either case we restrict our attention to $N_1 \times B$. In case (i) we define a coloring χ' on N_1 by setting $\chi'(n)$ equal to the constant $\chi(n, b)$. We find $N_2 \subseteq N_1$, N_2 infinite, so that χ' is either constant or distinct on it. Then χ is either monochromatic or column on $N_2 \times B$.

Case (ii) is slightly more complex. We define a coloring χ' on $[N_1]^2$ by

$$\chi'(\{n_1, n_2\}_<) = \{(b_1, b_2): b_1, b_2 \in B, \chi(n_1, b_1) = \chi(n_2, b_2)\} .$$

There are precisely $2^{|B|^2}$ colors for χ'. By Ramsey's theorem there exists $N_2 \subset N_1$ so that χ' is constant on $[N_2]^2$.

Suppose that, for some $n_1 < n_2 \in N_2$, $b_1 \neq b_2 \in B$, $\chi(n_1, b_1) = \chi(n_2, b_2)$. Then, for all $m < m' \in N_2$, $\chi(m, b_1) = \chi(m', b_2)$. There exists $n_3 \in N_2$, $n_3 > n_2$. Now $\chi(n_2, b_2) = \chi(n_1, b_1) = \chi(n_3, b_2) = \chi(n_2, b_1)$, which is impossible since all columns have distinct colors. Thus elements in distinct rows and distinct columns have distinct colors. Elements in distinct rows and the same columns have distinct colors by assumption, so all elements in distinct rows have distinct colors.

For each $b \in B$, if there exist $n_1 < n_2 \in N_2$ so that $\chi(n_1, b) = \chi(n_2, b)$, then $\chi(m, b) = \chi(m', b)$ for all $m < m' \in N_2$; that is, χ is either constant or distinct on each row. We find $B_1 \subset B$, $|B_1| = r+1$ so that either χ is constant on each row $b \in B_1$ or χ is distinct on each row $b \in B_1$. In the latter case, χ is distinct on $N_2 \times B_1$. In the former case, the row constants are distinct, since single columns are distinct, so that χ is row on $N_2 \times B_1$.

We give a simple coloring that provides a counterexample if $2r^2+1$ is replaced by $2r^2$. Decompose $[2r^2] = S_1 + \cdots + S_r + T_1 + \cdots + T_r$, where all $|S_i| = |T_j| = r$. If $s, s' \in S_i$ then (m, s) and (n, s') are colored identically for all m, n. If $t, t' \in T_j$ then (m, t) and (m, t') are colored identically for all m. Otherwise, all colors are distinct.

A Canonical Bipartite Ramsey theorem, analogous to Section 5.1, Theorem 6, can also be given. We leave the statement and the proof of this theorem to our readers.

Taylor [1976] generalizes Hindman's theorem in the same way in which Erdös and Rado generalized Ramsey's theorem to a Canonical Ramsey theorem. Taylor's theorem, indeed, could be called a Canonical Hindman theorem.

Let χ be a finite coloration of $[\omega]^{<\omega}$, the finite subsets of N. Then Hindman's theorem states that there exists a disjoint collection $\mathscr{D}$ such that $FU(\mathscr{D})$ is monochromatic. Now let us call a coloration χ of $FU(\mathscr{D})$ *canonical* if one of the following holds:

(i) χ is constant on $FU(\mathscr{D})$.
(ii) $\chi(E_1) = \chi(E_2)$ iff $\min(E_1) = \min(E_2)$.
(iii) $\chi(E_1) = \chi(E_2)$ iff $\max(E_1) = \max(E_2)$.
(iv) $\chi(E_1) = \chi(E_2)$ iff $\min(E_1) = \min(E_2)$ and $\max(E_1) = \max(E_2)$.
(v) $\chi(E_1) = \chi(E_2)$ iff $E_1 = E_2$.

Theorem 5. Let χ be a coloration (not necessarily finite) of $[\omega]^{<\omega}$. Then there exists a disjoint collection $\mathscr{D}$ such that χ is canonical on $FU(\mathscr{D})$.

The proof is rather difficult and is not given in this book.

One is struck by a similarity between many of the Canonical Ramsey theorems, a similarity that extends to the infinite case of Ramsey's theorem itself. Let S be a (possibly ordered) countably infinite set, and $F(S)$ be a family of structures defined on S. For example, $F(S)$ could be $[S]^2$ or $S \times S$. Call a coloring of $F(S)$ invariant if, under that coloring, every finite $T \subset S$ has $F(T)$ colored equivalently. Here two colorings are considered equivalent if they are identical under a bijection between S and T (preserving any ordering) and a bijection of the color names. Many of these results state that if χ is a coloring of $F(S)$, perhaps with some restrictions (as, e.g., limiting the number of colors to r), then there exists an infinite $T \subset S$ so that $F(T)$ is colored invariantly. We do not know whether one can make a general statement of this type.

5.6   EUCLIDEAN RAMSEY THEORY

In a series of papers, Erdös, Graham, Montgomery, Rothschild, Spencer, and Straus [1973, 1975a, 1975b] have examined a variety of problems that meld Ramsey theory to the geometry of Euclidean n-space R^n. Let K be a finite configuration in Euclidean space. We define a relation $R(K, n, r)$:

$R(K, n, r)$: Under any r-coloring of the points of R^n there exists a monochromatic $K' \cong K$. (Here $\cong$ is "congruence.")

For example, let K be an equilateral triangle of unit side (more precisely, the vertices of the triangle). We can color R^2 in strips of width $\sqrt{3}/2$, that is,

$$\chi(x, y) = \begin{cases} \text{red if } [2y/\sqrt{3}] \text{ is even}, \\ \text{blue if } [2y/\sqrt{3}] \text{ is odd}. \end{cases}$$

A simple geometry argument shows that no $K' \cong K$ is monochromatic so that $R(K, 2, 2)$ is false. If we 2-color R^4 and consider five points on an equilateral simplex of unit length, some three of these points must be the same color. Hence $R(K, 4, 2)$ is true.

One may replace "congruence" by other notions such as "similar" or "translate of." Let H be any group of symmetries of Euclidean space. We define:

$R_H(K, n, r)$: For any r-coloring of the vertices of R^n there exists a monochromatic K' and $\sigma \in H$ such that $\sigma K = K'$.

Gallai's theorem, given in Section 2.3, states that if H is the homothety group (and hence also the larger similarity group) then $R_H(K, n, r)$ holds for all finite configurations K in Euclidean n-space.

DEFINITION. K is Ramsey if, for all r, there exists n' so that, for $n \geqslant n'$, $R(K, n, r)$ holds.

The main question is the determination of the Ramsey configurations. Let K consist of two points at distance d. For all r, R' contains a simplex of $r + 1$ points, all at distance d. Any r-coloring yields a monochromatic K. Thus $R(K, r, r)$ holds for all r, and hence K is Ramsey. More generally, if K is an m-point equilateral simplex then $R(K, (m - 1)r, r)$ holds, so K is Ramsey.

Notation. Let $x = (x_1, \ldots, x_n) \in R^n$, $y = (y_1, \ldots, y_m) \in R^m$, Define

$$x * y = (x_1, \ldots, x_n, y_1, \ldots, y_m) \in R^{n+m}.$$

If $K_1 \subset R^n$, $K_2 \subset R^m$ define $K_1 * K_2 \subset R^{n+m}$ by

$$K_1 * K_2 = \{x * y : x \in K_1, y \in K_2\}.$$

Theorem 1. If K_1 and K_2 are Ramsey then $K_1 * K_2$ is Ramsey.

Proof. Fix $K_1 \subseteq R^n$, Ramsey, $K_2 \subseteq R^m$, Ramsey, and $r > 0$. Fix u so that $R(K_1, u, r)$ holds. By the Compactness principle there exists a finite $T \subseteq R^u$ so that any r-coloring of T yields a monochromatic K_1. Let $t = |T|$ and $T = \{x_1, \ldots, x_t\}$. Let v be such that $R(K_2, v, r')$ holds. We claim that $R(K_1 * K_2, u + v, r)$ holds. Let χ be an r-coloring of R^{u+v}. Define an r'-coloring χ' of R^v by

$$\chi'(y) = (\chi(x_1 * y), \ldots, \chi(x_t * y)) .$$

We find $K_2 \subseteq R^v$ monochromatic over χ'. Define an r-coloring χ'' of T by

$$\chi''(x_i) = \chi(x_i * y) \qquad \text{for any } y \in K_2 .$$

Let K_1 be monochromatic under χ''. Then $K_1 * K_2$ is monochromatic under χ.

Corollary 2. All bricks are Ramsey.

By a "brick" we mean a rectangular parallelopiped, that is, a set of the form $\{(x_1, \ldots, x_n): x_i = 0 \text{ or } a_i, 1 \le i \le n\}$. Clearly, any brick is a $*$-product of n two-point configurations.

Corollary 3. All subsets of bricks are Ramsey.

Clearly, if K is Ramsey all $K' \subset K$ are Ramsey. This corollary includes equilateral m-simplexes. In fact, it gives all known Ramsey configurations.

In view of the results given above it is tempting to conjecture that *all* finite sets K are Ramsey. This, however, is false. Let $K = \{0, 1, 2\}$, three points equally spaced on a line. We show that $R(K, n, 4)$ is false for all n by giving an explicit 4-coloring of R^n. Color $u \in R^n$ by $[|u|^2]$ (modulo 4), that is, for $0 \le i < 4$

$$\chi(u) = i \qquad \text{if } |u|^2 = 4a + i + \Theta, \quad a, i \in Z, \quad 0 \le \Theta < 1 .$$

Suppose that $\{x, y, z\} \cong K$. If we let y be the middle point there exists u, $|u| = 1$, so that $x = y + u$, $z = y - u$. Then

$$|x|^2 + |z|^2 = 2|y|^2 + 2|u|^2 = 2|y|^2 + 2 .$$

If $\chi(x) = \chi(y) = \chi(z) = i$ there exist $a_1, a_2, a_3 \in Z$, $0 \le \Theta_1, \Theta_2, \Theta_3 < 1$

with

$$4a_1 + i + \Theta_1 + 4a_2 + i + \Theta_2 = 2(4a_3 + i + \Theta_3) + 2$$

so that

$$\Theta_1 + \Theta_2 - 2\Theta_3 = 4(3a_3 - a_1 - a_2) + 2,$$

which is impossible. Thus K is not Ramsey.

We extend this example, using results on nonhomogeneous linear systems.

Theorem 4. Let $K = \{v_0, v_1, \ldots, v_k\}$ be such that there exist $c_0, \ldots, c_k, b \in R, b \neq 0$, satisfying

(*) $$\sum_{i=1}^{k} c_i(v_i - v_0) = 0,$$

(**) $$\sum_{i=1}^{k} c_i(|v_i|^2 - |v_0|^2) = b \neq 0.$$

Then K is not Ramsey.

Proof (outline). Let χ' be the coloring of R, given by Chapter 3, Theorem 23, such that the equation

$$\sum_{i=1}^{k} c_i(x_i - x_0) = b$$

has no solution with $\chi'(x_i) = \chi'(x_0)$ for $1 \leqslant i \leqslant k$. Define χ on R^n by

$$\chi(u) = \chi'(|u|^2).$$

One can show that any $K' = \{v_0', \ldots, v_k'\} \cong K$ still satisfies (*) and (**) [by showing that the system (*), (**) is preserved under rotation around origin and translation] so that K' cannot be monochromatic under χ.

The conditions of Theorem 4 are equivalent to saying that K is not spherical; that is, the vertices of K cannot be placed on a common sphere. We omit the proof. Note that all bricks are indeed spherical.

Summary. If K is a subset of a brick it is Ramsey
If K is not spherical it is not Ramsey.

Since publication of the first edition of this volume additional Ramsey configurations have been found. Let $k < s$ and let $x_1, \ldots, x_k$ be nonzero reals and let S be the set of points in R^s with precisely k nonzero coordinates having values $x_1, \ldots, x_k$ in that order. For example, with $k = 2$, $s = 3$, $x_1 = 1$, $x_2 = -1$, $S = \{(1, -1, 0), (1, 0, -1), (0, 1, -1)\}$ is an obtuse triangle.

Claim. S is Ramsey.

Let the number of colors r be arbitrary. Let n satisfy

$$n \rightarrow (s)_r^k$$

For each k-set $A \subset [n]$ let x_A denote the point with nonzero coordinates $x_1, \ldots, x_k$ in positions A. An r-coloring of the points x_A induces an r-coloring of $[n]^k$. A monochromatic s-set B of coordinates gives a monochromatic S.

Combining this result with Theorem 1, Frankl and Rödl [1986] have greatly extended the class of known Ramsey configurations. In particular, they have shown that all triangles are Ramsey. A complete characterization of Ramsey configurations has remained elusive.

Many of the finite questions $R(K, n, r)$ are very interesting. For example, it is conjectured that $R(K, 2, 2)$ is true for all K, $|K| = 3$, except the equilateral triangle. For many particular K's this is known to be true.

Finally, we explore Euclidean Ramsey questions where H is the group of translations. The results are negative. Let us call a set T achromatic if no two points of T are colored identically.

Theorem 5. Let $c = \binom{k}{2} + 1$. For all k-element sets $S \subseteq R$ there exists a c-coloring of R so that all translates $S + x$ are achromatic.

Proof. We first c-color any finite set $Y \subseteq R$ so that no translate of S contains two points with the same color. We color the points of Y in ascending order. To color $y \in Y$ we note that y lies on a common translate with at most $\binom{k}{2}$ previously colored points of Y [the points $y + (x' - x'')$ with $x', x'' \in S$, $x' < x''$] and so may be given a distinct coloring.

The Compactness principle implies that there exists a c-coloring of R. Note that the application of the Compactness principle renders this proof essentially nonconstructive.

Corollary 6. Let $c = \binom{k}{2} + 1$. For all k-element configurations $S =$

$\{v_1, \ldots, v_k\} \subseteq R^n$, with n-arbitrary, there exists a c-coloring of R^n so that all translates $S + x$ are achromatic.

Proof. Select a coordinate system so that the points of S have distinct first coordinates $x_1, \ldots, x_k$ By Theorem 5 there is a c-coloring of R so that all translates of $x_1, \ldots, x_k$ are achromatic. We color points of R^k by the colors of their first coordinates.

5.7 GRAPH RAMSEY THEORY

Graph Ramsey theory has grown from nonexistence 20 years ago to become one of the presently most active areas in Ramsey theory. Rather than attempt to present an encyclopedic collection of the wealth of results currently available, we will instead (following our usual philosophy) discuss a selection of those that we believe illustrates the variety of questions considered and techniques used in this area.

A major impetus behind the early development of Graph Ramsey theory was the hope that it would eventually lead to methods for determining larger values of the classical Ramsey numbers $R(m, n)$. However, as so often happens in mathematics, this hope has not been realized; rather, the field has blossomed into a discipline of its own. In fact, it is probably safe to say that the results arising from Graph Ramsey theory will prove to be more valuable and interesting than knowing the exact value of $R(5, 5)$ [or even $R(m, n)$].

The idea behind Graph Ramsey theory is basically as follows. For an arbitrary (fixed) graph G, we would like to determine the smallest integer $r = r(G)$ so that, no matter how the edges of K_r are 2-colored, a monochromatic subgraph isomorphic to G is always formed. For the classical Ramsey numbers, G itself is taken to be a complete graph. When k colors are used instead of two, we will denote the corresponding value of r by $r(G; k)$.

Just as in the classical case, it is convenient to consider the more general "off-diagonal" situation. For graphs $G_1, G_2, \ldots, G_k$, we let $r(G_1, G_2, \ldots, G_k)$ denote the least integer r so that, no matter how the edges of K_r are k-colored, for some i a copy of G_i occurs in the ith color. Of course, the existence of $r(G_1, G_2, \ldots, G_k)$ is guaranteed by Ramsey's original theorem.

To begin with, one of the simplest and most general results in Graph Ramsey theory is the following: For a graph G (which we will always assume has no isolated vertices), let $\chi(G)$ denote the chromatic number of G and let $c(G)$ denote the cardinality of the largest connected component of G.

Theorem 1 (Chvátal–Harary [1972]).

$$r(G, H) \geqslant (\chi(G) - 1)(c(H) - 1) + 1 .\tag{1}$$

Proof. Let $m = (\chi(G) - 1)(c(H) - 1)$, and consider K_m to be made up of $\chi(G) - 1$ copies of $K_{c(H)-1}$ with edges interconnecting all pairs of vertices in the different copies of $K_{c(H)-1}$. Color all the edges within a copy of $K_{c(H)-1}$ with color 2 and all remaining (interconnecting) edges with color 1. Certainly, there is no copy of G with color 1 since, if there were, we could color the vertices of this copy of G with $\chi(G) - 1$ colors, corresponding to the copies of K_{n-1} they lie in, and this contradicts the definition of $\chi(G)$. On the other hand, no copy of H with color 2 can occur, since the largest connected component in K_m with color 2 has $c(H) - 1$ vertices.

Theorem 1 can be applied to yield one of the most elegant results of Graph Ramsey theory, due to Chvátal [1977].

Theorem 2. For any tree T_m with m vertices

$$r(T_m, K_n) = (m - 1)(n - 1) + 1 .\tag{2}$$

Proof. The lower bound follows from (1). It remains to show that

$$r(T_m, K_n) \leqslant (m - 1)(n - 1) + 1 .\tag{3}$$

For $m = 2$ or $n = 2$, (3) is immediate. Assume that (3) holds for all values of m' and n' with $m' + n' < m + n$. Consider a 2-colored $K_{(m-1)(n-1)+1}$, using the colors red and blue, say. Let T' be a tree formed from T by the removal of some endpoint x (where x is connected by y in T). By the induction hypothesis, this $K_{(m-1)(n-1)+1}$ contains either a *blue* K_n (and we are done) or a *red* T'. Thus we may assume that there is a red T' in $K_{(m-1)(n-1)+1}$. We remove the $m - 1$ points of this red T', leaving a 2-colored $K_{(m-1)(n-2)+1}$. Again by the induction hypothesis, this graph contains either a red T or a blue K_{n-1}; we may clearly assume the latter.

Consequently, in the original $K_{(m-1)(n-1)+1}$ we have a red T' and a blue K_{n-1} disjoint from it. Finally, we examine the edges emanating from y to the blue K_{n-1}. If any of these edges is red then we have a red T. If not then all these edges are blue and there is a blue K_n. This completes the induction step, and the proof is finished.

There are still relatively few exact nontrivial values known for $r(G, H)$. One of the more interesting is the following.

Theorem 3 (Burr [1974]). Let T_m be a tree with m vertices, and assume that $m - 1$ divides $n - 1$. Then

$$r(T_m, K_{1,n}) = m + n - 1.$$ (4)

Proof. We first show that

$$r(T_m, K_{1,n}) \geq m + n - 1.$$ (5)

Let $k = (n - 1)/(m - 1)$. Form a 2-coloring of K_{m+n-2} by taking $k + 1$ copies of K_{m-1} (all having all red edges) interconnected by all blue edges. No red T_m has been formed, since T_m has m vertices. Also, no blue $K_{1,n}$ has been formed, since the largest blue degree in the K_{m+n-2} is $k(m - 1) = n - 1$. This proves (5).

Next, we show that

$$r(T_m, K_{1,n}) \leq m + n - 1.$$ (6)

For $m = 2$, (6) is immediate. As in the proof of Theorem 2, we form the tree T' by removing an endpoint x of T_m (which we assume is connected to y in T_m). In a 2-colored K_{m+n-1} we can assume by induction that there is either a blue $K_{1,n}$ (in which case we are done) or a red T'. We may assume the latter. Since there are $m + n - 1 - (m - 1) = n$ vertices v_i of K_{m+n-1} that are not vertices of the red T' and K_{m+n-1} contains no blue $K_{1,n}$, *some* edge from y to some v_i must be red. But this forms a red T in K_{m+n-1}. Thus by induction (6) holds, and the proof is complete.

The corresponding results when $m - 1$ does not divide $n - 1$ are much more complicated and are not completely understood. However, in this case (6) still holds; in fact, for almost all trees T_m,

$$r(T_m, K_{1,n}) = m + n - 2$$

for n sufficiently large.

A special case that has received particular attention is the one in which G consists of a number of disjoint copies of a particular graph. A particularly nice result of this type is due to Burr, Erdös, and Spencer [1975] for the graph nK_3 consisting of n disjoint triangles.

Theorem 4

$$r(nK_3) = 5n \qquad \text{for } n \geq 2.$$ (7)

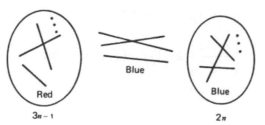

Figure 5.3

Proof. Recall that, for $n = 1$, we already know that $r(K_3) = 6$. To see that $r(nK_3) \geq 5n$, consider the 2-coloring of K_{5n-1} shown in Fig. 5.3. It is easily verified that this 2-coloring of K_{5n-1} contains no monochromatic nK_3.

We next show by induction on n that

$$r(nK_3) \leq 5n \qquad \text{for } n \geq 2. \tag{8}$$

The case $n = 2$ requires a detailed case analysis, which we omit. Fix a 2-coloring of K_{5n}, $n \geq 3$. We may find $\{(5n-5)/3\}$ vertex disjoint monochromatic triangles by merely selecting triangles until fewer than six points remain. As $\{(5n-5)/3\} \geq n$ we may assume that at least one of these triangles is in each color, say $\{1, 2, 3\}$ is colored red and $\{4, 5, 6\}$ is colored blue.

By a "bow tie" we mean a pair of triangles of different colors that share a common vertex (see Fig. 5.4). Assume, by symmetry, that at least five of the edges between $\{1, 2, 3\}$ and $\{4, 5, 6\}$ are blue. Then at least one of $1, 2, 3$, say 1, is joined by blue edges to at least two of $4, 5, 6$, say 4 and 5, and $\{1, 2, 3, 4, 5\}$ forms a bow tie. When these five points are deleted, there exists, by induction, a monochromatic $(n-1)K_3$ on the remaining $5n - 5$ points. This, together with the appropriate K_3 in the bow tie, gives a monochromatic nK_3.

The bow tie argument may be generalized to yield asymptotic results on $r(nG)$ and $r(mG, nG)$ for arbitrary G.

We next sample a few results when the number k of color classes is allowed to be larger than 2. Detailed proofs, unless otherwise attributed, can be found in Erdős and Graham [1975].

Figure 5.4 A bow tie.

Theorem 5. For a tree T_m with m edges

$$\left\lceil \frac{m-1}{2} \right\rceil (k-1) \le r(T_m; k) \le 2mk + 1. \tag{9}$$

Proof. To prove the lower bound (9), let t denote $[(m-1)/2]$. Consider $K_{t(k-1)}$ as $k-1$ copies of K_t, labeled as $K_t^{(1)}, K_t^{(2)}, \ldots, K_t^{(k-1)}$. That is, we can think of $K_{t(k-1)}$ as a complete graph of K_{k-1} with "fat" vertices $K_t^{(i)}$, $i = 1, 2, \ldots, k-1$. Now decompose the edge set of K_{k-1} into $k-1$ matchings. Then, if the edge $\{u \cdot v\}$ is in the ith matching, give all the edges in the complete bipartite graph between the fat vertices $K_t^{(u)}$ and $K_t^{(v)}$ the color i. Further, color all the edges in each of the $K_t^{(i)}$ with the color 0. It is now easy to see that this is a k-coloring of the edges of $K_{t(k-1)}$ with no monochromatic copy of $T(m)$ and the left-hand side of (9) follows.

If the conjecture of P. Erdös and V. T. Sós) (see Erdös [1964b]) that any T_m always occurs as a subgraph of any graph with d vertices and $[\frac{1}{2}(m-1)d] + 1$ edges were known to hold, then the upper bound of (9) could be strengthened to

$$r(T_m; k) < (m-1)k + 4 \tag{9'}$$

for k sufficiently large. It may well be that the right-hand side of (9') gives the correct asymptotic growth of $r(T_m; k)$. One reason for believing this is given by the following result.

Theorem 6. For a tree T_m with m edges

$$r(T_m; k) > (m-1)k - m^2$$

for k sufficiently large.

Proof. For a given large k, let k_0 denote the largest integer not exceeding k which is congruent to 1 modulo m. By a deep result of Ray-Chaudhuri and Wilson [1973], there will always exist a *resolvable* balanced incomplete block design $D_{k_0, m}$ having $(m-1)k_0 + 1$ points and $k_0(k_0 m - k_0 + 1)/m$ blocks of size m, provided only that k_0 is sufficiently large. Identify the points of $D_{k_0, n}$ with vertices of $K_{(m-1)k_0+1}$. Assign the color i to all edges of $K_{(m-1)k_0+1}$ that correspond to a pair of points occurring in the ith parallel class of $D_{k,n}$. This is a k_0-coloring of $K_{(m-1)k_0+1}$, which contains no monochromatic connected subgraph with $m+1$ vertices. Since $k_0 > k - m$, this induces a k-coloring of $K_{(m-1)k-m^2}$ with no monochromatic T_m.

For special trees, much more exact results are known. For example, for P_3, the path with three edges, Irving [1974] has shown that the following theorem holds.

Theorem 7

$$r(P_3; k) = \begin{cases} 2k+2 & \text{if } k \equiv 1 \ (\text{mod } 3), \\ 2k+1 & \text{if } k \equiv 2 \ (\text{mod } 3), \\ 2k \text{ or } 2k+1 & \text{if } k \equiv 0 \ (\text{mod } 3). \end{cases} \tag{10}$$

The proof of (10) also relies on the results for resolvable balanced incomplete block designs. In fact, it is not uncommon for the exact values of many of the known graph Ramsey numbers to depend on the existence (or nonexistence) of certain special combinatorial designs or algebraic structures. We see this, for example, in the determination of the classical Ramsey numbers $R(3, 3, 3)$ and $R(4, 4)$.

It is instructive to compare the Ramsey numbers of P_3 and the apparently closely related graph C_4, the cycle of four vertices. In fact, the Ramsey numbers for C_4 grow much more rapidly than those for P_3.

Theorem 8 (Chung–Graham [1975])

$$r(C_4; k) \leqslant k^2 + k + 2 \qquad \text{for all } k. \tag{11}$$

If $k - 1$ is a prime power then

$$r(C_4; k) \geqslant k^2 - k + 2. \tag{12}$$

Proof. We first need an estimate for the maximum number e of edges that a graph G with n vertices can have if G contains no C_4 as a subgraph. Let $A = (a_{ij})$ denote the adjacency matrix of such a G. Since $C_4 \not\subseteq G$,

$$\sum_{j=1}^n a_{i,j} a_{i',j} \leqslant 1 \tag{13}$$

for any choice of $1 \leqslant i < i' \leqslant n$. If c_j denotes $\sum_{i=1}^n a_{ij}$ then, summing (13) over all choices of i and i', we obtain

$$\sum_{j=1}^{n} c_j(c_j - 1) \leqslant n(n-1). \tag{14}$$

Applying the Cauchy-Schwarz inequality to (14), we have

$$2e = \sum_{j=1}^{n} c_j \leqslant \frac{n}{2} + n\sqrt{n - \frac{3}{4}}, \tag{15}$$

which is the sought after bound.

Suppose now that the edges of K_{k^2+k+2} are k-colored. Then at least one of the colors occurs on at least $(1/k)\binom{k^2+k+2}{2}$ edges. However, if we take

$$n = k^2 + k + 2, \qquad e \geqslant \frac{1}{k}\binom{k^2+k+2}{2}$$

we find that (15) is (barely) violated. This proves (11).

To prove (12), it is well known that since $k-1$ is a prime power there exists a simple difference set $D = \{d_1, \ldots, d_k\}$ (modulo $k^2 - k + 1$). For each t, $1 \leqslant t \leqslant k$, we form a cyclic (symmetric) matrix $B_t = (b_t(i, j))$ as follows:

$$b_t(i, j) = \begin{cases} 1 & \text{if } i + j + d_t \equiv d_s \pmod{k^2 - k + 1} \qquad \text{for some } d_s \in D, \\ 0 & \text{otherwise}. \end{cases}$$

Since D is a difference set, it follows that, for any choice of $i, j \in Z_{k^2-k+1}$, there exists t such that $b_t(i, j) = 1$. Furthermore, for each t no two rows of B_t have a common pair of 1's.

We now form a k-colored K_{k^2-k+1} as follows. The vertices of the K_{k^2-k+1} will be the elements of Z_{k^2-k+1}. For $i, j \in Z_{k^2-k+1}$, we color the edge $\{i, j\}$ with color t, where t is the least integer such that $b_t(i, j) = 1$. It follows from the preceding remarks that this is a k-coloring of K_{k^2-k+1} having no monochromatic C_4, and (12) is proved.

Again, note the appearance of combinatorial structures (in this case difference sets) in the lower bound proof. It follows from (11) and (12) and the fact that the prime powers are sufficiently dense that

$$r(C_4; k) \sim k^2$$

as $k \to \infty$.

One might be tempted to guess that, since C_3 is smaller than C_4, it should in some sense occur more readily in a k-coloring of a complete graph, and hence $r(C_3; k)$ would be substantially smaller than $r(C_4; k)$. In fact, however, exactly the opposite is true, as the following result shows.

Theorem 9

$$2^k < r(C_3; k) \leqslant 3k! \tag{16}$$

Proof. The lower bound in (16) follows easily by induction on k. When $k = 1$, it is certainly valid. If there exists a k-coloring of K_{2^k} with no monochromatic C_3 then, by joining two such copies of K_{2^k} by edges of color $k + 1$, we have a $(k + 1)$-colored $K_{2^{k+1}}$ with no monochromatic C_3.

To prove the upper bound, suppose that the edges of $K_{3k!}$ are arbitrarily k-colored. Some vertex v has at least $3(k - 1)!$ edges of the same color, say color k, joining v to vertices $x_1, \ldots, x_{3(k-1)!}$. If the complete graph spanned by the x_i's had an edge with color k, we would be done. Hence the x_i's must form a $(k - 1)$-colored $K_{3(k-1)!}$. By an induction assumption (which we did not actually make), this $K_{3(k-1)!}$ must have a monochromatic K_3. The proof is completed by noting that the right-hand side of (16) holds for $k = 1$ (to start the induction).

The reason behind the enormous difference between the growth rates of $r(C_4; k)$ and $r(C_3; k)$ is the fact that the much stronger Density theorem holds for C_4 but not for C_3. In fact, we have seen from (15) that if G has m vertices and $\frac{1}{2}m^{3/2} + \frac{1}{4}m$ edges then G must already contain C_4 as a subgraph. In contrast to this, as Turán's theorem shows, G can have essentially $m^2/4$ edges without containing a C_3.

Of course, the difference between even and odd cycles occurs frequently in Graph theory. It should not be surprising to find it appearing here as well.

For larger even cycles, the following is known.

Theorem 10. For any k and m,

$$r(C_{2m}; k) > (k - 1)(m - 1) .$$

If $k \leqslant 10^m/201m$ then

$$r(C_{2m}; k) \leqslant 201km .$$

Finally, there exist $\alpha > 0$ and a positive function g such that, for any $\varepsilon > 0$,

$$\alpha k^{1+1/2m} < r(C_{2m}; k) < g(m, \varepsilon)k^{1+[(1+\varepsilon)/(m-1)]} .$$

What this result says is that $r(C_{2m}; k)$ grows linearly in k out to a rather large value, and thereafter grows roughly like a power (exceeding 1) of k.

For odd cycles, the following analogue of Theorem 9 holds.

Theorem 11. For any k and m,

$$2^k m < r(C_{2m+1}; k) < 2(k + 2)!m .$$

An old question of Erdös asks whether or not, for some A, it is true that

$$r(C_3; k) \overset{?}{<} A^k .$$

It is not known whether

$$r(C_5; k) > r(C_3; k)$$

or even whether

$$r(C_{2m+1}; k) > r(C_3; k)$$

for m fixed and k large. In fact, Erdös has suggested that it might actually be easier to show that

$$r(C_{2m+1}; k) < A^k$$

for some $m > 1$ (especially if it is true!).

Anyone who would like to study more slowly growing Ramsey numbers is naturally led to the consideration of $r(F; k)$, where F is a forest (= acyclic graph) of some type. Some information is available in this case, although the known results are far from complete.

Theorem 12. If F_m is a forest with m edges then

$$\frac{k(\sqrt{m} - 1)}{2} < r(F_m; k) < 4km .$$

Furthermore, if $k \leq m^2$ then

$$r(F_m; k) > \alpha\sqrt{k}m$$

for a suitable positive constant α.

We omit the proof. For every forest F, $r(F; k)$ grows essentially linearly in k for large k [similarly to the behavior of $r(T; k)$ for trees T over the whole range k]. However, the second bound allows for the possibility that $r(F; k)$ can grow only as fast as $\sqrt{k}$ for small k. The next result shows that this can really happen.

Theorem 13. For a suitable constant α if $k \leq m$ then

$$r(mK_{1,m}; k) \leq \alpha\sqrt{k}m^2 .$$

Also, if $k \geq 3m^2$ then

$$r(mK_{1,m}; k) \leq 3km .$$

Note that, except for values of k between m and m^2, the upper and lower bounds for $r(mK_{1,m}; k)$ differ just by a constant factor.

Instead of letting $k \to \infty$, we might ask how slowly $r(G)$ can grow as the number of vertices of G grows. Here fairly precise results are available.

Theorem 14 (Burr–Erdös [1976]). If G is connected graph with m vertices then

$$r(G) \geq \left\lceil \frac{4m-1}{3} \right\rceil .$$

Furthermore, for each $m \geq 3$ there is such a G for which the bound is achieved.

If G is allowed to be disconnected (still without isolated vertices, though) then $r(G)$ can be smaller, as the following result shows.

Theorem 15 (Burr–Erdös [1976]). For a suitable positive constant α, if G has m vertices then

$$r(G) \geq m + \frac{\log m}{\log 2} - \alpha \log\log m .$$

Also, there is a graph G with m vertices and a constant β such that

$$r(G) \leqslant m + \beta\sqrt{m} .$$

Again, it is conjectured that the lower bound is essentially the truth.

We conclude this section with an assortment of results, conjectures, and remarks that suggest the variety of directions researchers in this field are currently pursuing. Most of these deal with the two-color case; the analogous statements for the k-color cases will be left to the reader.

To begin with, let us call a set $\mathcal{G} = \{G_1, G_2, \ldots\}$ of graphs an L-set if there is a constant $\alpha = \alpha(\mathcal{G})$ such that

$$r(G_i) \leqslant \alpha p(G_i)$$

for all i [where we recall that $p(G_i)$ denotes the number of vertices of G_i]. We define the edge density $\rho(G)$ of a graph G by

$$\rho(G) = \max_{H \subseteq G} \frac{e(H)}{p(H)} ,$$

where $e(H)$ denotes the number of edges of H. The following strong conjecture is due to P. Erdös.

Conjecture. If $\rho(G)$ is bounded for $G \in \mathcal{G}$, then $\mathcal{G}$ is an L-set.

Although the conjecture is far from being established at present, some supporting results are known; for example, any set of trees is an L-set, and for any m the set $\{C_i^m\}$ of mth powers of cycles is an L-set. A set of particular interest that is not yet known to be an L-set is the set $\{Q_i\}$ of cubes.

A very general result that is particularly useful for relatively dense graphs is the following.

Theorem 16 (Chvátal–Harary). Let G be a graph with p vertices and q edges, and let s be the order of the automorphism group of G. Then

$$r(G; k) \geqslant (sk^{q-1})^{1/p} . \tag{17}$$

Proof. The proof will be accomplished by a simple but effective use of the "probabilistic method." Let us (arbitrarily) label the vertices of G by $v_1, \ldots, v_p$, forming the labeled graph $\bar{G}$. It is easily seen that the complete graph K_n contains exactly

$(n)_p = n(n-1)\ldots(n-p+1)$ distinct copies of $\bar{G}$.

Thus, from the definition of s as the number of symmetries of G, K_n contains $t = (n)_p/s$ copies of G, say $G_1, G_2, \ldots, G_t$. Let us say that a k-coloring of K_n is G_i-*bad* if all the edges in the subgraph G_i of K_n have been assigned the same color. Since there are $k^{\binom{n}{2}}$ k-colorings altogether, there are just $k^{\binom{n}{2}-q+1}$ G_i-bad colorings. Therefore there are altogether at most $tk^{\binom{n}{2}-q+1}$ colorings that are G_i-bad for some i. Hence, if

$$tk^{\binom{n}{2}-q+1} < k^{\binom{n}{2}},$$

then *some* k-coloring must form no monochromatic copy of G in K_n. This condition certainly holds if

$$k^{q-1} > \frac{n^p}{s} \geq \frac{(n)_p}{s} = t,$$

that is,

$$n < (sk^{q-1})^{1/p}.$$

Thus

$$r(G; k) \geq (sk^{q-1})^{1/p},$$

and the theorem is proved.

An interesting variation is to determine for a given graph G the minimum number of edges a graph H can have so that any 2-coloring of the edges of H always results in a monochromatic copy of G. Let us denote this minimum number of edges by $r_e(G)$. It is obvious that

$$r_e(G) \leq \binom{r(G)}{2}. \tag{18}$$

When G is a complete graph then in fact, equality holds in (18) (see Erdös, Faudree, Rousseau, and Schelp [1978]). It is not known whether this can happen for noncomplete graphs. In the other direction it is not hard to show that

$$r_e(K_{1,m}) = \begin{cases} 2m - 1 & \text{for } m \text{ even}, \\ 2m & \text{for } m \text{ odd}. \end{cases}$$

Thus

$$\frac{r_e(K_{1,m})}{\binom{r(K_{1,m})}{2}} \to 0 \qquad \text{as } m \to \infty. \tag{19}$$

One of the most interesting open questions dealing with r_e is whether (19) holds when $K_{1,m}$ is replaced by the path P_m. At present we can rule out neither $r_e(P_m) < cm$ nor $r_e(P_m) > cm^2$ as a possibility.

Finally, one can consider the class $\mathscr{C}(G)$ of graphs H for which $H \to (G, G)$ but such that for any proper subgraph $H' \subset H$, $H' \not\to (G, G)$. Such graphs are called *Ramsey-minimal* for G by Burr, Faudree, and Schelp [1977]. It is known (see also Nešetřil and Rödl [1978b]) that $\mathscr{C}(G)$ is infinite in any of the following cases:

   (i) G is 3-connected,
  (ii) G has a chromatic number of at least 3,
 (iii) G is a forest that is not a union of stars.

The proof of (iii) is particularly illuminating since it incorporates several ideas that recur constantly in Graph Ramsey theory.

Theorem 17 (Nešetřil–Rödl [1978b]). Suppose that G is forest that is not a union of stars. Then $\mathscr{C}(G)$ is infinite.

Proof. What we show is that, for any given integer t, there is a graph $H \in \mathscr{C}(G)$ that has more than t vertices. Let n denote the number of vertices of G.

To begin with, we know by a classic result of Erdös [1959] that there exists a graph K with chromatic number $\chi(K)$ exceeding n^2 and girth exceeding t. In any 2-coloring of the edges of K, the edges of at least one of the colors form a graph K' with $\chi(K') > n$. [More generally, if $E(K) = E(K_1) \cup E(K_2)$ then $\chi(K) \le \chi(K_1)\chi(K_2)$, since the product of the two vertex colorings that achieve $\chi(K_1)$ and $\chi(K_2)$, respectively, gives a valid $\chi(K_1)\chi(K_2)$-coloring of the vertices of K.] By sequentially removing edges from K', we can form a *minimal* subgraph $K'' \subseteq K'$ with $\chi(K'') = n + 1$. In particular, all the vertices of K'' must have degree at least n, [If v is a vertex of K'' with degree less than n, then, by minimality, $K'' - \{v\}$ can be n-colored, and since v is adjacent to at most $n - 1$ of the vertices of $K'' - \{v\}$, this n-coloring can be extended to a valid n-coloring of K'' (which is impossible).] Furthermore, it is easy to see that K'' contains *every* forest F with n vertices as a subgraph (we simply start embedding F anywhere in K''; since all degrees in K'' are at least n, we never get "stuck").

Thus we have shown that $K \to G$. Therefore K contains a (minimal) subgraph $K^* \in \mathscr{C}(G)$. Note that K^* itself cannot be a forest since the edges of any forest can always be 2-colored so that no monochromatic path P_3 of three edges is formed (and by assumption, since G is not a union of stars, it contains P_3 as a subgraph). Thus the girth of K^* is finite; of course, it is greater than t, the girth of K. But this implies that K^* has more than t vertices, and the proof is complete.

In closing this section we mention a striking conjecture of Erdös.

Conjecture. If G_m has chromatic number m then

$$r(G_m) \geq r(K_m).$$

6

Beyond Combinatorics

6.1 TOPOLOGICAL DYNAMICS

In this section we outline the applications of topological dynamics to Ramsey theory. We shall prove van der Waerden's theorem and Hindman's theorem by these methods. We shall also show the implication of Szemerédi's theorem from the Ergodic theorem of Furstenberg.

We assume a rudimentary knowledge of topology. Let us review the product topology in the form we will require. Let B be a topological space, A a set. Set

$$X = \{f; A \rightarrow B\} .$$

We shall often write $X = B^A$; X forms a topological space under the product topology. For every $a_1, \ldots, a_s \in A$, $U_1, \ldots, U_s \subseteq B$ open,

$$U = \{f \in X: f(a_i) \in U_i, 1 \leq i \leq s\}$$

is an open set, and these sets form a basis for the product topology. When $B = [c]$, with the discrete topology, a basis for the open sets about $f \in X$ is given by these sets:

$$U = \{g \in X: f(a_i) = g(a_i), 1 \leq i \leq s\} .$$

When (B, ρ) is a metric space a basis is given by the sets:

$$U = \{g \in X: g(a_i) \in B(f(a_i), \varepsilon), 1 \leq i \leq s\} ,$$

where $\varepsilon > 0$ and $B(z, \varepsilon)$ denotes a ball of radius ε about z.

Let B be a compact space. (In our example B will be either $[c]$ or a compact metric space.) By the Tychonoff theorem $X = B^A$ is compact under the product topology. This property is central to all applications.

We observe that the Tychonoff theorem requires the Axiom of Choice so that all results will be nonconstructive.

We begin with construction of a topological space. Fix a number of colors $c \geq 2$. A c-coloring of Z is a function

$$x: Z \to [c] .$$

Let X denote the set of all such colorings; X is called the *bisequence space*, as its elements may be represented by doubly infinite sequences

$$x = (\cdots, x(-1), x(0), x(1), \cdots) .$$

We place on X the metric ρ, given by

$$\rho(x, y) = \begin{cases} (n+1)^{-1} & \text{if } n \geq 0 \text{ is minimal so that} \\ & x(i) = y(i) \quad \text{for } |i| < n , \\ 0 & \text{if } x = y . \end{cases}$$

If $x(0) \neq y(0)$ then $n = 0$ and $\rho(x, y) = 1$. The distance $\rho(x, y)$ is small iff x, y are identical near the origin. We may easily show that $x_n \to x$, under the metric ρ, iff $x_n(i) \to x_n(i)$ for all $i \in Z$. Topologically, $X = [c]^Z$ forms a compact space by the Tychonoff theorem. (Alternatively, any sequence $\{x_n\}$ contains a convergent subsequence by a diagonalization process.) Let $T: A \to X$ be defined by

$$(Tx)(i) = x(i+1) ,$$

where T is called the *shift operator*, and Tx is the coloring x moved one space to the left. T is bijective. It is uniformly continuous, as $\rho(x, y) < (n+1)^{-1}$ implies that $\rho(x, y) < n^{-1}$. Hence T is a homeomorphism (though it does not preserve the metric). For $s \in Z$, let T^s denote, as usual, the sth iterate of T, given by

$$(T^s x)(i) = x(i+s) .$$

Let $x \in X$. Define the orbital closure of x, denote by $\bar{x}$, by

$$\bar{x} = cl\{T^s x: s \in Z\} ,$$

where cl presents topological closure. Then $\bar{x}$ is a compact subspace of X. As $T(T^s x) = T^{s+1} x \in \bar{x}$, $T(\bar{x}) \subseteq \bar{x}$ by continuity. Similarly, $T^{-1}(\bar{x}) \subseteq \bar{x}$; thus T acts bijectively on $\bar{x}$.

DEFINITION. (Y, T) is a *dynamical system* if Y is a compact metric space and $T: Y \to Y$ is a bijective homeomorphism.

Theorem 1 (Topological Van Der Waerden Theorem). Let (Y, T) be a dynamical system, $r \geqslant 1$, $\varepsilon > 0$. Then there exist $y \in Y$, $n > 0$ so that

$$\rho(T^{in}y, y) < \varepsilon \qquad (\rho = \text{metric of } Y)$$

simultaneously for $1 \leqslant i \leqslant r$.

Theorem 2. Theorem 1 implies van der Waerden's theorem.

Proof. Let $x \in X$ be a c-coloring of Z. Apply Theorem 1 with $Y = \bar{x}$. For some $y \in \bar{x}$, $n > 0$

$$\rho(T^{in}y, y) < 1, \qquad 1 \leqslant i \leqslant r,$$

that is,

$$y(0) = T^n y(0) = \cdots = T^{nr} y(0)$$

or

$$y(0) = y(n) = \cdots = y(rn) .$$

As $y \in \bar{x}$, there exists $s \in Z$, $\rho(y, T^s x) < (rn + 1)^{-1}$, that is, y and $T^s x$ are identical on $[-rn, +rn]$ so that

$$T^s x(0) = T^s x(n) = \cdots = T^s(rn)$$

or

$$x(s) = x(s + n) = \cdots = x(s + rn) ,$$

a monochromatic arithmetic progression of length r. We have shown that an arbitrary finite coloring of Z contains arbitrarily long monochromatic APs. The replacement of Z by N is a simple exercise (either using the Compactness principle or considering symmetric colorings of Z).

DEFINITION. Let $x \in X$. A sequence of length r of x is an ordered r-tuple $(x(i), \ldots, x(i + r - 1))$, $i \in Z$. Let $\text{Seq}(x)$ denote the family of all sequences of a coloration x.

Theorem 3. $y \in \bar{x}$ iff $\text{Seq}(y) \subseteq \text{Seq}(x)$.

Proof. Let $y \in \bar{x}$, $(y(i), \ldots, y(i+r-1)) \in \text{Seq}(y)$. For some $s \in Z$

$$\rho(T^s x, y) < [1 + \max(|i|, |i+r-1|)]^{-1}$$

so that y and $T^s x$ agree on $[i, i+r-1]$ and

$$(y(i), \ldots, y(i+r-1)) = (x(i+s), \ldots, x(i+s+r-1)) \in \text{Seq}(x).$$

Conversely, assume that $\text{Seq}(y) \subseteq \text{Seq}(x)$, and let $n \geq 1$ be arbitrary. Then

$$(y(-n), \ldots, y(n)) \in \text{Seq}(x)$$

so, for some s,

$$(y(-n), \ldots, y(n)) = (x(-n+s), \ldots, x(n+s))$$

and $\rho(y, T^s x) < (n+1)^{-1}$. As n was arbitrary, $y \in \bar{x}$.

Theorem 4. The following conditions on $x \in X$ are equivalent:

- (i) $\text{Seq}(x)$ is minimal in the sense that, for no $y \in X$, is $\text{Seq}(y) \subset \text{Seq}(x)$.
- (ii) $y \in \bar{x} \Rightarrow \bar{y} = \bar{x}$.
- (iii) $\bar{x}$ is minimal in the sense that, for no $y \in X$, is $\bar{y} \subset \bar{x}$.
- (iv) (Bounded Gaps condition) For every $(x(i), \ldots, x(i+r-1)) \in \text{Seq}(x)$ there exists M so that, for every $t \in Z$, there exists $s \in [t, t+M-r]$ so that

$$(x(i), \ldots, x(i+r-1)) = (x(s), x(s+1), \ldots, x(s+r-1)),$$

that is, the sequence occurs as a subsequence of every interval of length M.

We call $x \in X$ *minimal* if conditions (i)–(iv) hold. Conditions (i) and (iv) apply when X is the bisequence space; (ii) and (iii) apply for any compact space X and homeomorphism T.

Proof. The equivalences (i) $\Leftrightarrow$ (ii) $\Leftrightarrow$ (iii) are immediate from Theorm 3. Let x satisfy (iv) and $y \in \bar{x}$. Let $(x(i), \ldots, x(i+s-1)) \in \mathrm{Seq}(x)$ with M given by (iv). Then $(y(1), \ldots, y(M)) \in \mathrm{Seq}(x)$ so

$$(y(1), \ldots, y(M)) = (x(t), \ldots, x(t+M-1))$$

for some $t \in Z$. For some $s \in [t, t+M-r]$

$$
\begin{aligned}
(x(i), \ldots, x(i+r-1)) &= (x(s), x(s+1), \ldots, x(s+r-1)) \\
&= (y(s-t+1), \ldots, y(s-t+r)) \\
&\in \mathrm{Seq}(y) .
\end{aligned}
$$

Hence (iv) $\Rightarrow$ (ii).

Conversely, let $s - (x(i), \ldots, x(i+r-1)) \in \mathrm{Seq}(x)$, and let t_M be defined for all odd (for convenience) $M = 2N+1$ so that s is not a subsequence of x on $[t_M, t_M + 2N]$. Let $x_M = T^{t_M+N}x$ so that s is not a subsequence of x_M on $[-N, N]$. By the Compactness principle (i.e., Diagonal argument) there exists a convergent subsequence $x_{M_i} \to y$, $M_i = 2N_i + 1$. As $x_{M_i} \in \bar{x}$, $y \in \bar{x}$. For every N there exists i so that $N_i \geqslant N$ and x_{M_i} is identical with y on $[-N, N]$. Thus y does not contain s as a subsequence on $[-N, N]$. As N is arbitrary, $s \notin \mathrm{Seq}(y)$ so x does not satisfy (ii).

Theorem 5 (Minimal Property). For every $x \in X$ there exists $y \in \bar{x}$, y minimal.

Proof. We use condition (iii). Let $\mathcal{U} = \{\bar{y}: y \in X\}$. Suppose that $\mathcal{C} \subseteq \mathcal{U}$ forms a chain under containment. Any finite subfamily $\bar{y}_1, \ldots, \bar{y}_n \in \mathcal{C}$ has a minimal set, so $\cap \bar{y}_i \neq \emptyset$. $\mathcal{C}$ is a family of closed sets of X. As X satisfies the Finite Intersection property (equivalent to Compactness),

$$\bigcap_{\bar{y} \in \mathcal{C}} \bar{y} \neq \emptyset ,$$

that is, there exists $z \in \bar{y}$ for all $\bar{y} \in \mathcal{C}$. Thus $\bar{z} \subseteq \bar{y}$ for all $\bar{y} \in \mathcal{C}$.

In $\mathcal{U}$ every chain $\mathcal{C}$ is dominated (under containment) by some $\bar{z} \in \mathcal{U}$. By Zorn's lemma (logically equivalent to the Axiom of Choice) every $\bar{x} \in \mathcal{U}$ is dominated by some minimal element $\bar{y} \subseteq \bar{x}$.

The Minimal property has a combinational interpretation. Let $\mathcal{A}$ be a set of sequences. Suppose that one wishes to show that, for all colorings $x \in X$, $\text{Seq}(x) \cap \mathcal{A} \neq \emptyset$. Then, using condition (i), it suffices to show that, for all minimal colorings $x \in X$, $\text{Seq}(x) \cap \mathcal{A} \neq \emptyset$. Observe that van der Waerden's theorem is a statement of this type. The minimal colorings, unfortunately, may have a complicated structure. We note, for example, that, for all real α and $n \in N$, the coloring

$$x(i) = [\alpha i] \text{ (modulo } n)$$

gives a minimal coloring. When α is irrational, x is not periodic.

We call a dynamical system (Y, T) minimal if $\bar{y} = Y$ for all $y \in Y$. By the Minimal property, for all dynamical systems (Y, T) there exists a nonempty $A \subseteq Y$ so that (A, T) is minimal.

Proof of Theorem 1. By the above remarks, we may assume that (Y, T) is a minimal dynamical system. Consider the following sequence of statements:

$$
\begin{aligned}
&(A_r): &&\forall \varepsilon > 0\, \exists x, y, n \quad \rho(T^{in}x, y) < \varepsilon, &&1 \leq i \leq r \\
&(B_r'): &&\forall \varepsilon > 0\, \forall z \exists x, n \quad \rho(T^{in}x, z) < \varepsilon, &&1 \leq i \leq r \\
&(B_r): &&\forall \varepsilon > 0\, \forall z \exists x, n, \varepsilon' > 0, && \\
&&&T^{in}[B(x, \varepsilon') \subseteq (z, \varepsilon), &&1 \leq i \leq r \\
&(C_r): &&\forall \varepsilon > 0\, \exists w, n \quad \rho(T^{in}w, w) < \varepsilon, &&1 \leq i \leq r
\end{aligned}
$$

We shall show that $(A_r) \Rightarrow (B_r) \Rightarrow (B_r') \Rightarrow (C_r) \Rightarrow (A_{r+1})$. Observe that (A_1) is trivial, taking x, n arbitrary, $y = T^n x$.

Theorem 6. Let (Y, T) be minimal. For all $\varepsilon > 0$ there exists $M > 0$ so that, for all $x, y \in Y$,

$$\min_{|s| \leq M} \rho(T^s x, y) < \varepsilon.$$

Proof. When Y is a set of colorations, the existence of M follows from the Bounded Gaps condition. In general, if no m existed, there would be sequences x_i, $y_i \in Y$ so that $\rho(T^s x_i, y_i) > \varepsilon$ for all $|s| \leq i$. On an appropriate subsequence, x_i, y_i would simultaneously converge to x, y and $\rho(T^s x, y) \geq \varepsilon$ for all $s \in Z$. But then $y \notin \bar{x}$, contradicting the minimality assumption.

$(A_r) \Rightarrow (B_r)$. Let $\varepsilon > 0$ be fixed and M satisfy Theorem 6. Let ε' be such that $\rho(x, y) < \varepsilon'$ implies $\rho(T^s x, u) < \varepsilon$ for $|s| \leq M$ (ε' exists by the uniform continuity of the T^s). Let x, y, n satisfy (A_r) for ε'. Let $z \in Y$ be arbitrary. For some s, $|s| \leq M$, $\rho(T^s y, z) < \varepsilon$. As $\rho(T^{in}x, y) < \varepsilon'$, $\rho(T^{in+s}x, T^s y) < \varepsilon$ so that $\rho(T^{in}x^*, z) < 2\varepsilon$, where $x^* = T^s x$. As ε was arbitrary, (B_r) follows.

$(B_r) \Rightarrow (B_r')$. Fix $\varepsilon > 0$, $z \in Y$. Select, by (B_r), x, n so that $T^{in}x \in B(z, \varepsilon/2)$, $1 \leq i \leq r$. Select, by continuity, an $\varepsilon' \leq \varepsilon$ such that $x' \in B(x, \varepsilon')$ implies $T^{in}x' \in B(T^{in}x, \varepsilon/2)$, $1 \leq i \leq r$. (Note that ε' depends strongly on n.) Then $x' \in B(x, \varepsilon')$ implies that

$$\rho(T^{in}x', z) < \rho(T^{in}x', T^{in}x) + \rho(T^{in}x, z) \leq \varepsilon,$$

as desired.

$(B_r') \Rightarrow (C_r)$. Let $\varepsilon_0 < \varepsilon/2$ and $z_0 \in Y$ be arbitrary. Let $z_1, n_1, \varepsilon_1 \leq \varepsilon_0$ satisfy

$$T^{in_1}[B(z_1, \varepsilon_1)] \subseteq B(z_0, \varepsilon_0), \qquad 1 \leq i \leq r.$$

By induction, select z_s, n_s, $\varepsilon_s \leq \varepsilon_{s-1}$ so that

$$T^{in_s}[B(z_s, \varepsilon_s)] \subseteq B(z_{s-1}, \varepsilon_{s-1}), \qquad 1 \leq i \leq r.$$

By the Compactness principle, from the infinite sequence $\{z_s\}$ we may find $t < s$ such that $\rho(z_s, z_t) < \varepsilon_0$. (The perspicacious reader will note here the focusing of progressions, as in the combinatorial proof of van der Waerden's theorem.) By a simple induction

$$T^{i(n_s + n_{s-1} + \cdots + n_{t+1})}[B(z_s, \varepsilon_s)] \subseteq B(z_t, \varepsilon_t)$$
$$\subseteq B(z_s, 2\varepsilon_0)$$
$$\subseteq B(z_s, \varepsilon).$$

Thus (C_r) is satisfied by $n = n_s + \cdots + n_{t+1}$ and $w = z_s$.

$C_r \Rightarrow A_{r+1}$. Let $\varepsilon > 0$ be arbitrary and w, n satisfy (C_r). Set $x = T^{-n}w$, $y = w$. For $1 \leq i \leq r+1$,

$$\rho(T^{in}x, y) = \rho(T^{(i-1)n}w, w) < \varepsilon,$$

completing the proof of Theorem 1.

For $S \subseteq Z$ let $P(S)$ denote the set of nonempty finite sums of S. Hindman's theorem states that if N is finitely colored there exists an infinite $S \subseteq N$ such that $\mathscr{P}(S)$ is monochromatic. Let (X, T) be a dynamical system.

DEFINITION. We say that $x, y \in X$ are *proximal* if

$$\inf_{n \in Z} \rho(T^n x, T^n y) = 0 .$$

If X is the bisequence space then $x, y \in X$ are proximal iff there are arbitrarily long intervals $I \subset Z$ such that $x(i) = y(i)$ for $i \in I$.

Theorem 7 (Topological Hindman's Theorem).   Let (X, T) be a dynamical system, $x \in X$, $\bar{x} = X$. Let $Y \subseteq X$ be minimal. Then there exists $y \in Y$ such that x, y are proximal.

Note that if X itself is minimal (e.g., $X =$ unit circle, $T =$ rotation by θ with $\theta / 2\pi$ irrational) then $Y = X$ so we may take $y = x$.

Theorem 8.   Theorem 7 implies Hindman's theorem.

Proof.   Let $x: N \rightarrow [c]$ be arbitrary. For technical reasons extend x to Z be setting $x(-i) = x(i)$, $i \in N$, $x(0)$ arbitrary. In the bisequence space set $X = \bar{x}$. Let Y be a minimal subset of X [which exists by the Minimal property (Theorem 5)]. Let $y \in Y$ be given by Theorem 7. We use two properties:

(i) x, y are proximal,

and

(ii) y has the Bounded Gaps property.

Either $\inf_{n \in N} \rho(T^n x, T^n y) = 0$ or $\inf_{n \in -N} \rho(T^n x, T^n y) = 0$. Assume the former. (One says that x, y are positively proximal.) Let $y(0) =$ red. For some M, every M-interval of y contains a red point. Let I be an M-interval, $I > 0$, on which x, y coincide. Then, for some $a_1 \in I$, $x(a_1) = y(a_1) =$ red. Now, by induction, assume $0 < a_1 < \cdots < a_s$ have been found. Set $u = a_1 + \cdots + a_s$. For some M, every M-interval of y contains a u-interval identical to $[0, u]$. More formally:

$$\exists_M \forall_n \exists_{m \in [n, n+M-u]} y(m + i) = y(i) , \qquad 0 \leq i \leq u .$$

There exists an M-interval $I > a_s$ on which x, y coincide. Set a_{s+1} equal to the minimal element of I. Thus

$$y(a_{s+1} + i) = x(a_{s+1} + i) , \qquad 0 \leq i \leq u .$$

Let $\beta \in \mathscr{P}(\{a_1, \ldots, a_s, a_{s+1}\})$. Either $\beta \in \mathscr{P}(\{a_1, \ldots, a_s\})$ or $\beta = a_{s+1} + \alpha$, where $\alpha \in \mathscr{P}(\{a_1, \ldots, a_s\})$, or $\beta = a_{s+1}$. In the first case

$$x(\beta) = \text{red}$$

by induction. In the second case

$$x(\beta) = x(a_{s+1} + \alpha) = y(a_{s+1} + \alpha) = y(\alpha) = \text{red}$$

by induction. In the third case

$$x(\beta) = x(a_{s+1}) = y(a_{s+1}) = y(0) = \text{red},$$

completing the induction step. The infinite set $S = \{a_1, a_2, \cdots\}$ is the desired infinite monochromatic set.

As $S \subseteq N$, $\mathscr{P}(S)$ is monochromatic under the original x. When $\inf_{n \in -N} \rho(T^n x, T^n y) = 0$, the above argument would yield an $S \subseteq -N$ such that $\mathscr{P}(S)$ is monochromatic. But then $-S \subseteq N$, and $\mathscr{P}(-S)$ would be monochromatic under the original coloring.

For completeness (and for the edification of topology buffs) we prove Theorem 7. Let (X, T) be a dynamical system. Let $X^x = \{f: X \to X\}$ with the product topology. As X is compact, X^x is compact. X^x forms a semigroup under composition. [Notation: $(fg)(x) = f(g(x))$.] The sets

$$\mathcal{O} = \{h \in X^x: \rho(h(x), y) < \varepsilon\}$$

form a subbasis for the topology. For any $g \in X^x$, the right multiplication $\Psi_g: X^x \to X^x$, given by $\Psi_g(f) = fg$, is continuous, for with $\mathcal{O}$ given above

$$\Psi_g^{-1}(\mathcal{O}) = \{f \in X^x: \rho(f(g(x)), y) < \varepsilon\}$$
$$= \{f \in X^x: \rho(f(x'), y) < \varepsilon\}, \qquad x' = g(x),$$

is open. The left multiplication $\Phi_g: X^x \to X^x$, given by $\Phi_g(f) = gf$, is continuous if g is, for in that case

$$\Psi_g^{-1}(\mathcal{O}) = \{f \in X^x: \rho(g(f(x)), y) < \varepsilon\}$$
$$= \{f \in X^x: f(x) \in g^{-1}(B(y, \varepsilon))\}$$

is open as $g^{-1}(B(y, \varepsilon))$ is. Set

$$E = cl\{T^n: n \in Z\} \subseteq X^x.$$

Then $f \in E$ iff, for all $x_1, \ldots, x_s \in X$, $\varepsilon > 0$, there exists $n \in N$ so that

$$\rho(f(x_i), T^n(x_i)) < \varepsilon, \qquad 1 \le i \le s$$

(the existence of $f \in E$ other than $f = T^n$ is nonconstructive). E is a closed subset of compact X^x, hence E is compact.

We claim that E is closed under composition. Set $f, g \in E$, and let $x_1, \ldots, x_s \in X$, $\varepsilon > 0$ be arbitrary. For some $n \in Z$

$$\rho(f(g(x_i)), T^n(g(x_i))) < \frac{\varepsilon}{2}, \qquad 1 \le i \le s.$$

There exists $\delta > 0$ so that $\rho(x, y) < \delta$ implies that $\rho(T^n x, T^n y) < \varepsilon/2$. For some $m \in Z$

$$\rho(g(x_i), T^m(x_i)) < \delta, \qquad 1 \le i \le s.$$

Hence

$$\rho(T^n(x_i), T^n T^m(x_i)) < \frac{\varepsilon}{2}, \qquad 1 \le i \le s,$$

and so

$$\rho(fg(x_i), T^{n+m}(x_i)) < \varepsilon, \qquad 1 \le i \le s.$$

Therefore $fg \in E$, and E is thus a semigroup. It is called the *enveloping semigroup* of (X, T).

Theorem 9 (Idempotent Theorem). Let E be a compact semigroup for which right multiplication $\Psi_g: E \to E$, given by $\Psi_g(f) = fg$, is continuous for all $g \in E$. Then there exists $g \in E$ such that $g^2 = g$.

Proof. Let $\mathcal{A}$ denote the family of compact semigroups $A \subseteq E$. $\mathcal{A} \ne \emptyset$ as $E \in \mathcal{A}$. If $\mathcal{C} \subseteq \mathcal{A}$ is a chain then $\cap \, \mathcal{C} \in \mathcal{A}$. ($\cap \, \mathcal{C} \ne \emptyset$ as all $A \in \mathcal{C}$ are compact.) By Zorn's lemma there exists a minimal $A \in \mathcal{A}$. Let $g \in A$. Then Ag is a semigroup $((f_1 g)(f_2 g) = (f_1 g f_2)g)$ and is compact by continuity. As $Ag \subseteq A$, $Ag = A$ by minimality. Set $B = \{f \in A: fg = g\}$. As $Ag = A$, $B \ne \emptyset$. B is a semigroup ($f_1 g = g$ and $f_2 g = g$ imply that $f_1 f_2 g = g$) and is compact by continuity. Thus $B = A$ by minimality. As $g \in B$, $g^2 = g$.

Proof of Theorem 7. Fix (X, T), $x \in X$ with $\bar{x} = X$ and $Y \subseteq X$ minimal. Let E be the enveloping semigroup of (X, T). Set

$$F = \{f \in E: fx \in Y\}.$$

We first show F is nonempty. Let $x' \in Y$ be arbitrary. As $x' \in \bar{x}$ there is a sequence n_m with $T^{n_m}x \to x'$. As E is compact the T^{n_m} cluster at some $f \in E$. Then $fx = x'$ so $f \in F$. F is closed, hence compact, from the topology of X^x.

Let $z \in Y$, $f \in E$. For all $\varepsilon > 0$ there exists n with

$$\rho(fz, T^n z) < \varepsilon .$$

But $T^n z \in Y$ so $\rho(fz, Y) < \varepsilon$. As ε was arbitrary and Y closed, $fz \in Y$, that is $f(Y) \subseteq Y$ for all $f \in E$. Now let $f_1, f_2 \in F$:

$$f_1 f_2 x = f_1(f_2 x) \in f_1(Y) \subseteq Y .$$

Hence $f_1 f_2 \in F$. We have shown that F is a compact semigroup. Let $g \in F$ be the idempotent guaranteed by Theorem 9.

We claim that gx is proximal to x. Note that $gx \in Y$ since $g \in F$. Let $\varepsilon > 0$. Since $g \in E$, there exists n so that

$$\rho(gx, T^n x) < \frac{\varepsilon}{2} ,$$

$$\rho(g(gx), T^n(gx)) < \frac{\varepsilon}{2} .$$

But $g(g(x)) = g(x)$. Thus

$$\rho(T^n x, T^n gx) < \varepsilon ,$$

completing the proof.

Let $S \subseteq Z$. We say that S has positive asymptotic density if, for some $\alpha > 0$, there is a sequence of intervals $[n_i, m_i) \subset Z$ such that $m_i - n_i \to \infty$ and

$$\lim \frac{|S \cap [n_i, m_i)|}{m_i - n_i} = \alpha .$$

Szemerédi's Theorem. If $S \subseteq Z$ has positive asymptotic density then it contains arithmetic progressions of length k for all k.

Furstenberg's Theorem. Let $(Y, \mathcal{U}, \mu)$ be a probability space and $T: X \to X$ a measure preserving bijection. For all $A \subseteq Y$ with $\mu(A) > 0$ and all k there exists n such that

$$\mu[A \cap T^n A \cap \cdots \cap T^{n(k-1)}A] > 0 .$$

The proof of Furstenberg's theorem involves recondite methods of Ergodic theory and will not be considered here. We shall only show the implication of Szemerédi's theorem from Furstenberg's theorem. Demonstration of the equivalence of this statement of Szemerédi's theorem to that of Section 2.5 (via the Compactness principle) is left to the reader.

A map $L: 2^Z \to [0, 1]$ is called a *norm* if $L(Z) = 1$ and $L(A \cup B) = L(A) + L(B)$ for all disjoint A, B. Also, L is called *shift invariant* if $L(T + i) = L(T)$ for all $T \subseteq Z$, $i \in Z$. We require the existence of a shift-invariant norm L (often called a Banach norm) with $L(S) = \alpha$.

Let $\mathcal{U} = \{f: 2^Z \to [0, 1]\}$. Under the product topology (giving $[0, 1]$ the usual topology) $\mathcal{U}$ is compact by the Tychonoff theorem. Let $\mathcal{L} \subseteq \mathcal{U}$ denote the set of all norms. We claim that $\mathcal{L}$ is a closed set. Let $L \in cl(\mathcal{L})$, and A, B be arbitrary disjoint sets. For all $\varepsilon > 0$ there exists $M \in \mathcal{L}$ so that

$$|L(X) - M(X)| < \varepsilon, \qquad X = A, B, A \cup B,$$

as $M(A) + M(B) = M(A \cup B)$,

$$|L(A \cup B) - L(A) - L(B)| < 3\varepsilon.$$

Since ε was arbitrary, $L(A \cup B) - L(A) - L(B) = 0$. Similarly, $L(Z) = 1$ so that $L \in \mathcal{L}$. Under the subset topology, $\mathcal{L}$ forms a compact space
Define $L_i \in \mathcal{L}$ by

$$L_i(X) = \frac{|X \cap [n_i, m_i)|}{m_i - n_i}.$$

By compactness, $\{L_i\}$ has an accumulation point L. For some subsequence i',

$$L(S) = \lim_{i'} L_{i'}(S).$$

As $\lim L_i(S) = \alpha$, $L(S) = \alpha$. Let $X \subseteq Z$ be arbitrary. For some subsequence i',

$$L(X) = \lim_{i'} L_{i'}(X),$$

$$L(X + 1) = \lim_{i'} L_{i'}(X).$$

For all i,

$$|L_i(X+1) - L_i(X)| = \frac{\|(X+1) \cap [n_i, m_i)| - |X \cap [n_i, m_i)\|}{m_i - n_i} \leq \frac{2}{m_i - n_i}.$$

Therefore

$$\lim_{i'} L_{i'}(X+1) - L_{i'}(X) = 0 \qquad \text{for every subsequence } i'.$$

Hence $L(X) = L(X+1)$, and L is the desired shift-invariant norm. Set $Y = \{0, 1\}^Z$. For $i \in Z$ set

$$Y_i = \{y \in Y: y(i) = 1\}.$$

Define μ by

$$\mu(Y_{i_1} \cap \cdots \cap Y_{i_s}) = L((S + i_1) \cap \cdots \cap (S + i_s)).$$

This generates a measure μ on the σ-algebra $\mathcal{U}$ generated by the Y_i. (Here μ is a probability distribution on the finite algebra generated by $Y_{-n}, \ldots, Y_n$ by the finite additivity of L. The extension of μ to $\mathcal{U}$ is given by the classic Kolmogoroff Extension theorem.)

Now we may begin. Let $T: Y \to Y$ be the shift operator, given by $(Ty)(i) = y(i+1)$. As L is shift invariant, T is measure preserving. Set $A = Y_0$ so that $\mu(A) = L(S) = \alpha$. Let $k > 0$ be arbitrary. By Furstenberg's theorem there exists n so that

$$0 < \mu[A \cap T^n A \cap \cdots \cap T^{(k-1)n}A]$$
$$= L[S \cap (S + n) \cap \cdots \cap (S + (k-1)n)].$$

There exists a so that

$$a \in S + in, \qquad 0 \leq i \leq k-1,$$

and $\{a, a - n, \ldots, a - (k-1)n\} \subseteq S$ is the desired AP for Furstenberg's theorem.

Note. We have followed the approach of Furstenberg and Weiss [1978]. Furstenberg [1977] and Szemerédi [1975] present full proofs of Szemerédi's theorem. Veech [1977] gives a survey of developments in topological dynamics. Bergelson and Hindman [in press] explore the connections between topological dynamics and Ramsey theory. The book of Furstenberg [1981] is highly recommended.

6.2   ULTRAFILTERS

An *ultrafilter* on a set X is a zero-one finitely additive measure μ defined on all subsets of X, that is,

  (i) $\mu(A) = 0$ or 1 for all $A \subset X$ and $\mu(X) = 1$;
  (ii) $\mu(A_1 \cup \cdots \cup A_n) = \mu(A_1) + \cdots + \mu(A_n)$ if the A_i are pairwise disjoint.

For all $i \in X$ the measure μ_i, given by

$$\mu_i(A) = \begin{cases} 1 & \text{if } i \in A, \\ 0 & \text{if } i \notin A, \end{cases}$$

is called a principal ultrafilter. If μ is not of this form, it is called a nonprincipal ultrafilter. Equivalently, a nonprincipal ultrafilter μ satisfies the condition

  (iii) $\mu(A) = 0$ if A is finite.

Alternatively, an ultrafilter can be described as a family $\mathscr{A} \subset 2^X$, satisfying the conditions

  (iv) $X \in \mathscr{A}, \varnothing \notin \mathscr{A}$;
  (v) for all $A \subset X$ either $A \in \mathscr{A}$ or $A^c \in \mathscr{A}$ [not both, by (iv) and (vi)];
  (vi) $A, B \in \mathscr{A}$ implies that $A \cap B \in \mathscr{A}$.
Also, for nonprincipal ultrafilters,
  (vii) A finite implies that $A \notin \mathscr{A}$.

The equivalence is readily seen by setting

$$\mathscr{A} = \{ A \subset X \colon \mu(A) = 1 \}.$$

Theorem 1.   There exist nonprincipal ultrafilters on any infinite set X.

Proof.   We call $\mathscr{B} \subset 2^X$ a *filter* if it satisfies conditions (iv) and (vi). Let $\mathscr{F}$ denote the set of filters. If $\mathscr{C} \subset \mathscr{F}$ is a chain under containment, then $\cup \mathscr{C}$ is a filter that contains all $\mathscr{A} \in \mathscr{C}$. By Zorn's lemma (which is required for this result) every filter is contained in a maximal filter. Let $\mathscr{A}$ be a maximal filter, $B \subset X$, $B \notin \mathscr{A}$. Then

$$\mathscr{A}^+ = \mathscr{A} \cup \{ A \cap B^c \colon A \in \mathscr{A} \}$$

is a filter, so, by maximality, $\mathscr{A}^+ = \mathscr{A}$ and $B^c \in \mathscr{A}$. In concise terms maximal filters are ultrafilters.

Now set $\mathscr{B} = \{A \subset X : X - A \text{ is finite}\}$. Since $\mathscr{B}$ is a filter (we require X to be infinite so that $\varnothing \notin \mathscr{B}$), it is contained in an ultrafilter $\mathscr{A}$. If A is finite, $X - A \notin \mathscr{B} \subset \mathscr{A}$ so that $A \notin \mathscr{A}$. Thus $\mathscr{A}$ is a nonprincipal ultrafilter.

We illustrate the use of ultrafilters with a proof of Ramsey's theorem.

Theorem 2. Let $\chi : [N]^2 \to [r]$ be arbitrary. There exists an infinite monochromatic $A \subset N$.

Proof. Fix a nonprincipal ultrafilter μ on $[N]^2$. Define $\chi' : N \to [r]$ by

$$\chi'(x) = \text{that } i \text{ so that } \mu(\{y : \chi(\{x, y\}) = i\}) = 1 \, .$$

As i ranges over $[r]$, the above sets partition $N - \{x\}$ (of unit measure since μ is nonprincipal) so that exactly one such i has this property. Now N is partitioned by χ' so that there exists a unique i such that

$$\mu(B) = 1 \, , \qquad \text{where } B = \{x : \chi'(x) = i\} \, .$$

We next find an infinite set of color i. Choose $a_1 \in B$ arbitrarily. Having chosen $a_1, \ldots, a_n$, set

$$S = B \cap \bigcap_{j=1}^{n} \{y : \chi(\{a_j, y\}) = i\} \, .$$

Since S is the intersection of $n + 1$ sets of unit measure, $\mu(S) = 1$. Choose $a_{n+1} \in S$, distinct from $a_1, \ldots, a_n$. Then $A = \{a_n : n \in N\}$ is the desired monochromatic set.

We have actually proved the existence of an infinite monochromatic set in the color most often used—where "most often" is in terms of the ultrafilter!

Our next result, noted by N. Hindman, gives a general connection between ultrafilters and Ramsey theory.

Theorem 3. Let $\mathscr{G}$ be a family of nonempty subsets of X. The following are equivalent:

(i) If X is finitely colored there exists a monochromatic $G \in \mathcal{G}$.

(ii) There exists an ultrafilter $\mathcal{A}$ on X such that, for all $A \in \mathcal{A}$, $A \supseteq G$ for some $G \in \mathcal{G}$.

Proof

(ii) $\Rightarrow$ (i). If X is finitely colored then the set A of points of some particular color is in the ultrafilter and the $G \in \mathcal{G}$ with $G \subset A$ is monochromatic.

(i) $\Rightarrow$ (ii). Set

$$\mathcal{B} = \{A \subset X: A \cap G \neq \emptyset \text{ for all } G \in \mathcal{G}\}.$$

Let $A_1, \ldots, A_k \in \mathcal{B}$ be arbitrary. Partition X into 2^k parts by the Venn diagram of the A_i. Some $G \in \mathcal{G}$ is contained in one part. Since $G \cap A_i \neq \emptyset$, $1 \leq i \leq k$, we must have $G \subset A_1 \cap \cdots \cap A_k$ so that $A_1 \cap \cdots \cap A_k \neq \emptyset$. Now let $\mathcal{B}^+$ be the set of finite intersections of sets in $\mathcal{B}$. Then $\mathcal{B}^+$ is a filter, so it is contained in an ultrafilter $\mathcal{A}$. If $A \in \mathcal{A}$ then $A^c \notin \mathcal{A}$, $A^c \in \mathcal{B}$, $A^c \cap G = \emptyset$ for some $G \in \mathcal{G}$, and thus $A \supseteq G$.

We now present Glazer's startling proof of Hindman's theorem (Chapter 3, Theorem 15). The product topology $\mathcal{T}$ on the set of all zero-one functions on 2^N forms a compact space by the Tychonoff theorem. Let $\mathcal{U}$ denote the set of ultrafilters over N. $\mathcal{U}$ is a closed subspace of $\mathcal{T}$. Since $\mathcal{T}$ is compact, $\mathcal{U}$ is compact under the product topology. The sets

$$\{\mu \in \mathcal{U}: \mu(A) = \varepsilon\}, \qquad \varepsilon = 0, 1, \quad A \subset N,$$

from a subbasis for the topology on $\mathcal{U}$.

We define a binary operation $+$ on $\mathcal{U}$ by

$$(\mu + \nu)(A) = \mu(\{n: \nu(A - n) = 1\}).$$

(By $A - n$ we mean $\{x \in N: x + n \in A\}$.) To show closure, let μ, ν be arbitary ultrafilters. Clearly $(\mu + \nu)(N) = 1$, $(\mu + \nu)(\emptyset) = 0$. Let $A, B \subset N$ be disjoint. Then $(A \cup B) - n = (A - n) \cup (B - n)$, and at most one of these disjoint sets may have unit measure under ν. Thus

$$\{n: \nu(A \cup B - n) = 1\} = \{n: \nu(A - n) = 1\} \cup \{n: \nu(B - n) = 1\},$$

a disjoint union. Hence

$$(\mu + \nu)(A \cup B) = \mu(\{n: \nu(A \cup B - n) = 1\})$$
$$= \mu(\{n: \nu(A - n) = 1\}) + \mu(\{n: \nu(B - n) = 1\})$$
$$= (\mu + \nu)(A) + (\mu + \nu)(B),$$

and so $\mu + \nu$ is an ultrafilter. The operation is associative, as

$$(\mu + (\nu + \sigma))(A) = ((\mu + \nu) + \sigma)(A)$$
$$= \mu(\{m: \nu(\{n: \sigma(A - n) = 1\}) = 1\}).$$

For fixed ν the right addition $\Psi_\nu: \mathcal{U} \to \mathcal{U}$, given by $\Psi_\nu(\mu) = \mu + \nu$, is continuous since

$$\{\mu: (\mu + \nu)(A) = \varepsilon\} = \{\mu: \mu(B) = \varepsilon\},$$

where $B = \{n: \nu(A - n) = 1\}$.

We apply the Idempotent theorem (Section 6.1, Theorem 9) to deduce the existence of an ultrafilter μ such that $\mu + \mu = \mu$. Since the principal ultrafilters satisfy $\mu_i + \mu_i = \mu_{2i} \neq \mu_i$ for all $i \in N$, μ is nonprincipal. Fix this μ.

With the existence of the appropriate ultrafilter established, the remainder of the proof is brief. Let N be finitely colored. The set of points colored some particular color has unit measure under μ; let A_0 denote that set. For any $B \subset N$ define $B^* = \{n: \mu(B - n) = 1\}$. If $\mu(B) = 1$ then

$$1 = (\mu + \mu)(B) = \mu(B^*) \quad \text{and} \quad \mu(B \cap B^*) = 1.$$

Select $a_1 \in A_0 \cap A_0^*$, and set $A_1 = A_0 \cap (A_0 - a_1) - \{a_1\}$ so that $A_1 \subset A_0$, $a_1 + A_1 \subset A_0$, and $\mu(A_1) = 1$. Having defined A_n of unit measure, select $a_{n+1} \in A_n \cap A_n^*$ and set $A_{n+1} = A_n \cap (A_n - a_{n+1}) - \{a_{n+1}\}$ so that $A_{n+1} \subset A_n$, $a_{n+1} + A_{n+1} \subset A_n$, and $\mu(A_{n+1}) = 1$. Then all sums of $X = \{a_n: n \in N\}$ are clearly the same color. Well, perhaps not so clearly; take $a_1 + a_3 + a_5$, for example. Since $a_5 \in A_4 \subset A_3$, $a_3 + a_5 \in A_2 \subset A_1$, $a_1 + a_3 + a_5 \in A_0$, as are all sums from X.

Note. Comfort [1977] provides an overview of ultrafilter methods.

6.3 AN UNPROVABLE THEOREM

We define a set $S \subseteq N$ to be *large* if $|S| > \min(S)$ (e.g., $\{3, 7, 56, 914\}$ is large, but $\{4, 7, 8\}$ is not). Let us modify the Ramsey arrow notation and write

$$m \underset{*}{\rightarrow} (n)^k_r \tag{1}$$

if for any r-coloring of $[n, m]^k$ there exists a large monochromatic set. We define a statement (PH), initially considered by J. Paris and L. Harrington:

$$\text{(PH)} \quad \forall_{n,k,r} \exists_m m \underset{*}{\rightarrow} (n)^k_r .$$

Let us prove (PH). Fix n, k, r. Let $\mathscr{A}$ be the family of finite large sets $S \subseteq [n, \infty)$. If $[n, \infty)^k$ is r-colored there exists, by Ramsey's theorem, an infintie monochromatic set $T \subseteq [n, \infty)$. Let S denote the first $\min(T)$ elements of T. Then $S \in \mathscr{A}$, and S is monochromatic. The existence of a finite m now follows directly from the Compactness principle (Section 1.5, especially Version C).

Theorem 1 (Paris–Harrington). In Peano arithmetic (PH) is unprovable.

Gödel's Incompleteness theorem implies the existence of statements about the integers that are true but unprovable in Peano arithmetic. The statement (PH) is the first natural example of such a statement.

Let $LR(n, k, r)$ denote the minimal m satisfying (1). [To simplify the presentation we shall deal with a modified statement (PH'), where "large" is replaced by $|X| \geq h(\min(X))$ for a function h. In fact, only technical modifications would be required for the Paris–Harrington theorem.] R. Solovay has shown that the function LR grows extremely rapidly. From a classic result in proof theory this implies that (PH) is unprovable in Peano arithmetic. We shall, for our modified (PH'), employ Solovay's methods. A lower bound on the analogous functions will be found by the construction of specific colorations. Our approach is self-contained, with the exception of the critical application of proof theory to show unprovability.

We first examine (PH) and LR for $k = 2$. Define $LR(n, r) = LR(n, 2, r)$. Here there will be no logical difficulties. The statement (PH) restricted to $k = 2$ (in fact, to any fixed k) is provable in Peano arithmetic. Technically these results are not necessary fot the Paris–Harrington theorem, but they are of interest in their own right and provide a "warm-up" for the general case.

We define an Ackermann hierarchy of functions:

$$f_1(x) = 2x ,$$

$$f_{n+1}(x) = f_n^{(x)}(x) ,$$

where $f^{(x)}$ denotes the xth iterate of f. This is a slight modification of the hierarchy defined in Section 2.7.

Theorem 2. $LR(n, r) \geqslant f_r(n)$.

Proof. We give an explicit r-coloring of $[n, f_r(n)]^2$. Let $n \leqslant x < y < f_r(n)$. We define $\overline{xy}$ to be the minimal i so that, for some j,

$$x, y \in [f_i^{(j)}(n), f_i^{(j+1)}(n)) .$$

We color $\{x, y\}$ by $\overline{xy}$. Observe that $\overline{xy} \leqslant r$ since we may take $i = r, j = 0$. Let $X = \{x_1, x_2, \ldots, x_m\}$ be a monochromatic set colored i. Then

$$\{x_1, \ldots, x_m\} \subseteq [s, f_i(s)) ,$$

where $s = f_i^{(j)}(n)$. If $i = 1$ then $m \leqslant f_1(s) - s = s$, so X is not large. For $i > 1$, $s = f_{i-1}^{(k)}(n)$ for some k and $f_i(s) = f_{i-1}^{(k+s)}(n)$. Thus

$$[s, f_i(s)) = \bigcup_{t=k}^{s+k-1} [f_{i-1}^{(t)}(n), f_{i-1}^{(t+1)}(n)) ,$$

a decomposition into s subintervals. Since all $\overline{x_u x_v} = i > i - 1$, the elements of X belong to distinct subintervals. Hence $m \leqslant s$, and X is not large.

Observe that the function $LR(n, n)$ grows at least as fast as the Ackermann function, hence faster than any primitive recursive function. In fact (though we do not show it here), $LR(n, n)$ may be bounded in both directions in terms of the Ackermann function.

We now outline some necessary prerequisites on ordinal numbers. Let $\gamma_1 = \omega$, $\gamma_2 = \omega^\omega$, $\gamma_{s+1} = \omega^{\gamma_s}$ for $s \geqslant 2$. Set $\varepsilon_0 = \lim \gamma_s$. As ordinal $\alpha < \varepsilon_0$ has a unique representation, called the Cantor normal form, as

$$\alpha = n_1 \omega^{\alpha_1} + n_2 \omega^{\alpha_2} + \cdots + n_t \omega^{\alpha_t} , \tag{2}$$

where $\alpha_1 > \alpha_2 > \cdots > \alpha_t \geqslant 0$, $n_i \in N$. If $\alpha < \gamma_{s+1}$ then all exponents $\alpha_i < \gamma_s$. For convenience we set $v_\beta(\alpha)$ equal to the coefficient of ω^β in α, $v_\beta(a) = 0$, if ω^β does not appear in the representation. In dealing with ordinals we use the standard notation

$$[\alpha] \stackrel{\text{def}}{=} \{\beta: \beta < \alpha\} .$$

DEFINITION. For $\alpha < \varepsilon_0$ we define

$$T(\alpha) = |\{\beta: v_\beta(\alpha) > 0\}| = \text{the number of terms of } \alpha$$
$$[= t \text{ if } \alpha \text{ is given by (2)}],$$
$$N(\alpha) = 1 + \text{the maximal integer to appear in the Cantor}$$
$$\text{normal form of } \alpha.$$

For example,

$$N[3\omega^{\omega^7 + 1} + 5\omega^{\omega + 4}] = 8 .$$

Technically, we define N inductively by $N(n) = n + 1$ for $n < \omega$ and

$$N(\alpha) = \max(n_1 + 1, \ldots, n_t + 1, N(\alpha_1), \ldots, N(\alpha_t))$$

for α given by (2).

Let $e_s(n)$ be defined inductively by $e_1(n) = n$, $e_{s+1}(n) = n^{e_s(n)}$, that is, $e_s(n)$ is a tower of n's of height s. The following property follows from a simple induction.

Property 1. If $\alpha < \gamma_{s+1}$, $T(\alpha) < e_s(N(\alpha))$.

Let $\alpha < \varepsilon_0$ be represented as in (2). Observe that α is a limit ordinal iff $\alpha_t > 0$. For every such limit ordinal we define a particular countable sequence approaching α with the nth term denoted by $\alpha(n)$ as follows.

CASE 1. α_t is not a limit ordinal. Write $\alpha_t = \beta_t + 1$. Set

$$\alpha(n) = n_1 \omega^{\alpha_1} + \cdots + n_{t-1}\omega^{\alpha_{t-1}} + (n_t - 1)\omega^{\alpha_t} + n\omega^{\beta_t} .$$

CASE 2. α_t is a limit ordinal. By induction $\alpha_t(n)$ has been defined. Set

$$\alpha(n) = n_1 \omega^{\alpha_1} + \cdots + n_{t-1}\omega^{\alpha_{t-1}} + (n_t - 1)\omega^{\alpha_t} + \omega^{\alpha_t(n)} .$$

These are the "natural sequences." Some examples are as follows:

$$\alpha = \omega^2 , \qquad \alpha(n) = n\omega ,$$
$$\alpha = \omega^2 + 3\omega , \qquad \alpha(n) = \omega^2 + 2\omega + n ,$$
$$\alpha = 5\omega^{\omega^2 + 2\omega} , \qquad \alpha(n) = 4\omega^{\omega^2 + 2\omega} + \omega^{\omega^2 + \omega + n} .$$

Property 2. $N(\alpha(n)) \leq \max[N(\alpha), n]$.

Again the proof is a simple induction. It may occur that $\alpha(n)$ is itself limit ordinal. Let α be a limit ordinal, $n \in N$. Consider the sequence $\alpha_0, \alpha_1, \ldots,$ defined by $\alpha_0 = \alpha$, $\alpha_{i+1} = \alpha_i(n)$, the sequence terminating when α_k is not a limit ordinal. As this is a decreasing sequence of ordinals, it must eventually terminate. We let $\alpha((n))$ denote the final term of the sequence. For example,

$$\omega^2((5)) = 4\omega + 5 \,,$$

$$\omega^{\omega^\omega}((3)) \text{ has 43 terms.}$$

Property 2'. $N(\alpha((n))) \leq \max[N(\alpha), n]$.

Now we extend the Ackermann hierarchy of functions. For $\alpha < \varepsilon_0$ we define $f_\alpha: N \to N$ inductively by

$$f_1(x) = 2x \,,$$

$$f_{\alpha+1}(x) = f_\alpha^{(x)}(x) \,,$$

$$f_\alpha(x) = f_{\alpha(x)}(x) \,, \qquad \alpha \text{ a limit ordinal} \,.$$

Equivalently, when α is a limit ordinal we may define

$$f_\alpha(x) = f_{\alpha((x))}(x) \,.$$

This transfinite sequence of functions is sometimes called the Grzegorczyk or Wainer hierarchy.

We say that a function f dominates a function g if, for some c, $f(n) > g(n)$ for all $n \geq c$. Let $P(s, t)$ be a two-variable statement such that

$$\forall_s \exists_t P(s, t) \tag{3}$$

is provable (and can be stated) in Peano arithmetic. Assume P is provably recursive; that is, there is an algorithm for deciding if $P(s, t)$ is true and a proof in Peano arithmetic that the algorithm always terminates. Let $f_P(s)$ denote the minimal t such that $P(s, t)$ holds. A classical proof theory result (and this is the only place where we require mathematical logic) states that f_P is dominated by f_α for some $\alpha < \varepsilon_0$.

We define $\varepsilon_0(n) = \gamma_n$ and extend the Ackermann hierarchy by defining

$$f_{\varepsilon_0}(n) = f_{\gamma_n}(n) \,.$$

Property 3. f_{ε_0} dominates all f_α, $\alpha < \varepsilon_0$.

Although this property appears obvious, the proof requires an examination of the limit sequences. We observe that if $\alpha < \beta < \varepsilon_0$, $N(\alpha) \leq m$, and β is a limit ordinal then $\alpha < \beta(m)$. We claim that if $\alpha < \beta < \varepsilon_0$. $2 \leqslant m$, and $N(\alpha) \leqslant m$ then $f_\alpha(m) < f_\beta(m)$. The proof uses transfinite induction on β. If β is not a limit ordinal, say $\beta = \delta + 1$, then

$$f_\beta(m) = f_\delta^{(m-1)}(f_\delta(m)) > f_\delta(m) > f_\alpha(m)$$

by induction. If β is a limit ordinal then $\alpha < \beta(m)$ by our observation so that $f_\beta(m) = f_{\beta(m)}(m) > f_\alpha(m)$ by induction. Hence we have shown that f_α is dominated by f_β whenever $\alpha < \beta < \varepsilon_0$. Finally, let $\alpha < \varepsilon_0$ be arbitrary, and s be such that $\alpha < \gamma_s$. For $m > \max(s, N(\alpha))$

$$f_{\varepsilon_0}(m) = f_{\gamma_m}(m) > f_\alpha(m)$$

so that f_{ε_0} dominates f_α.

Remark. Consider the mathematical parlor game of describing, on a single sheet of paper, as large an integer as you can. Extend the Ackermann hierarchy to $\alpha < \varepsilon_0 + \omega$ by the inductive definition. Now

$$\boxed{f_{\varepsilon_0+9}(9)}$$

should win against all nonlogicians. Indeed, by the proof theory results, the function f_{ε_0} lies "beyond the scope" of Peano arithmetic. We emphasize that these functions are recursive. There is a computer program that, given input n, computes—theoretically, of course!—$f_{\varepsilon_0}(n)$.

Let $P(s, t)$ be a statement such that

(PH0)	$P(s, t)$ is expressible in Peano arithmetic;
(PH1)	P is provably recursive;
(PH2)	$P(s, t)$ is false for all $t < f_{\varepsilon_0}(s)$;
(PH3)	for all s, $P(s, t)$ is true for some t.

Combining our previous remarks, statement (3) is a formula of Peano arithmetic that is unprovable in Peano arithmetic but true for the natural numbers. We shall construct a statement P of this form.

We begin by defining a sequence of colorings on ordinals. Our objective is to find colorings so that, if S is monochromatic, then $|S|$ may be bounded by $\max(S)$.

DEFINITION. Let $\beta < \alpha < \varepsilon_0$, and define

$$\overline{\alpha\beta} = \max\{\delta : v_\delta(\alpha) \neq v_\delta(\beta)\} .$$

Observe that $\delta = \overline{\alpha\beta}$ is well defined and $v_\delta(\alpha) > v_\delta(\beta)$.

Property 4. If $\alpha_1 > \cdots > \alpha_n$ then $\overline{\alpha_1 \alpha_n} = \max_{1 \leqslant i < n} \overline{\alpha_i \alpha_{i+1}}$.

We define a 3-coloring χ^* on $[\varepsilon_0]^3$ by

$$\chi^*(\{\alpha, \beta, \gamma\}_>) = \begin{cases} 0 & \text{if } \overline{\alpha\beta} > \overline{\beta\gamma}, \\ 1 & \text{if } \overline{\alpha\beta} = \overline{\beta\gamma}, \\ 2 & \text{if } \overline{\alpha\beta} < \overline{\beta\gamma}. \end{cases}$$

For the remainder of this section let S, $\alpha_1, \ldots, \alpha_r$, r satisfy the following:

$$S = \{\alpha_1, \ldots, \alpha_r\}, \qquad \alpha_1 = \max(S), \qquad (4)$$
$$\alpha_1 > \cdots > \alpha_r, \qquad r = |S|.$$

Property 5. If $\chi^*(S) = 1$ then $r \leqslant N(\alpha_1)$.

Proof. Set $\delta = \overline{\alpha_i \alpha_{j+1}}$, $1 \leqslant i < r$. Then $v_\delta(\alpha_1) > \cdots > v_\delta(\alpha_r)$ so $v_\delta(\alpha_1) \geqslant r - 1$ and $N(\alpha_1) \geqslant 1 + v_\delta(\alpha_1) \geqslant r$.

Property 6. If $\chi^*(S) = 2$ then $r \leqslant T(\alpha_1) + 1$. If, in addition, $\alpha < \gamma_{s+1}$ then $r \leqslant e_s(N(\alpha_1)) + 1$.

Proof. Let $\beta_i = \overline{\alpha_i \alpha_{i+1}}$, $1 \leqslant i < r$. By Property 4, $\beta_i = \overline{\alpha_1 \alpha_{i+1}}$. The β_i are increasing, hence distinct, and $v_{\beta_i}(\alpha_1) \neq 0$ so $r - 1 \leqslant T(\alpha_1)$. The second statement follows from Property 1.

Property 7. If $\chi^*(S) = 0$ and $\alpha < \omega^\omega$ then $r \leqslant N(\alpha_1) + 1$.

Proof. Let $\omega^s \leqslant \alpha < \omega^{s+1}$, $s < \omega$. Set $\beta_i = \overline{\alpha_i \alpha_{i+1}}$, $1 \leqslant i < r$, as before. Then $s \geqslant \beta_1 > \cdots > \beta_{r-1} \geqslant 0$ so $r \leqslant s + 2 \leqslant N(\alpha_1) + 1$.

Note. The assumption $\alpha < \omega^\omega$ is essential for Property 7. For example,

$$S = \{\omega^\omega, \omega^n, \omega^{n-1}, \ldots, \omega\}$$

has $\chi^*(S) = 0$, $\alpha = \omega^\omega$, and $|S|$ arbitrarily large.
We define, for $s \geqslant 2$, $(2s - 1)$-colorings

$$\chi_s: [\gamma_s]^{s+1} \to \{0, 1, 2, \ldots, 2s - 3, 2s - 2\}$$

and monotone functions $h_s: N \to N$ such that, if S is monochromatic under χ_s, then $r \leqslant h_s(N(\alpha_1))$. For $s = 2$

$$\chi_2: [\omega^\omega]^3 \to \{0, 1, 2\}$$

is the restriction of χ^*. By Properties 5, 6, and 7 we may take $h_2(x) = x + 1$. We define χ_3 in detail before proceeding to the inductive step.

Let $T = \{\alpha_1, \alpha_2, \alpha_3, \alpha_4\}_> \in [\omega^\omega]^4$. If $\chi^*(\{\alpha_1, \alpha_2, \alpha_3\}) = 1$ or 2 set $\chi_3(T) = 3$ or 4 (the "new" colors), respectively. Otherwise set $\alpha'_i = \overline{\alpha_i \alpha_{i+1}}$, $1 \leq i < 4$. If $\alpha'_1 > \alpha'_2 > \alpha'_3$ set

$$\chi_3(T) = \chi_2(\{\alpha'_1, \alpha'_2, \alpha'_3\})$$

(observe that $\alpha_i < \gamma_3$ implies $\alpha'_i < \gamma_2$ so that this is well defined). If not, set $\chi_2(T) = 0$ (actually, anything but 3 or 4 will do).

Let S, given by (4), be monochromatic under χ_3. If $\chi_3(S) = 3$ and 4 then $S - \{\alpha_r\}$ is 1 or 2 under χ^* so

$$r - 1 \leq N(\alpha_1) \qquad \text{or} \qquad r - 1 \leq e_2(N(\alpha_1)) + 1 \,,$$

respectively, by Property 5 or 6. Assume that S is another color, and set $\alpha'_i = \overline{\alpha_i \alpha_{i+1}}$, $1 \leq i < r$. For $i \leq r - 3$, $\chi^*(\{\alpha_i, \alpha_{i+1}, \alpha_{i+2}, \alpha_{i+3}\}) = 0$ so $\alpha'_i > \alpha'_{i+1}$, that is, $\alpha'_1 > \cdots > \alpha'_{r-2}$. Let $i_1 < i_2 < i_3 \leq r - 2$ be arbitrary. Then

$$\chi_2(\{\alpha'_{i_1}, \alpha'_{i_2}, \alpha'_{i_3}\}) = \chi_3(\{\alpha_{i_1}, \alpha_{i_2}, \alpha_{i_3}, \alpha_{i_3+1}\})$$

(i.e., the α' "mirror" the α). Hence $\{\alpha'_1, \ldots, \alpha'_{r-2}\}$ is monochromatic under χ_2 so

$$r - 2 \leq h_2(N(\alpha'_1)) \leq h_2(N(\alpha_1))$$

[as h_2 is monotone and $N(\alpha'_1) \leq N(\alpha_1)$]. Setting

$$h_3(x) = \max[x + 1, e_2(x) + 2, h(x) + 2] \,,$$

we have $r \leq h_3(N(\alpha_1))$ in all cases.

Now we give the general inductive step. Assume that χ_t, h_t have been defined for $2 \leq t < s$. Let

$$T = \{\alpha_1, \ldots, \alpha_{s+1}\}_> \in [\gamma_s]^{s+1} \,.$$

If $\chi^*(\{\alpha_1, \alpha_2, \alpha_3\}) = 1$ or 2 set $\chi_s(T) = 2s - 3$ or $2s - 2$ (the new colors), respectively. Otherwise, set $\alpha'_i = \overline{\alpha_i \alpha_{i+1}}$, $1 \leq i \leq s$. If $\alpha'_1 > \cdots > \alpha'_s$ set $\chi_s(T) = \chi_{s-1}(\{\alpha'_1, \ldots, \alpha'_s\})$. (This is well defined as all $\alpha'_i < \gamma_{s-1}$.) If not, set $\chi_s(T) = 0$. (Actually, anything but $2s - 3$ or $2s - 2$ will do.)

Let S, given by (4), be monochromatic under χ_s. If $\chi_s(S) = 2s - 3$ or $2s - 2$ then $S - \{\alpha_{r-s+3}, \ldots, \alpha_r\}$ is 1 or 2 under χ^* [as every triple is the initial three elements of an $\{s + 1\}$-set of S] so

$$r - s + 2 \leqslant N(\alpha_1) \qquad \text{or} \qquad r - s + 2 \leqslant e_{s-1}(N(\alpha_1)) + 1$$

by Property 5 or 6, respectively. Assume that S is some other color, and set $\alpha_i' = \overline{\alpha_i \alpha_{i+1}}$, $1 \leqslant i < r$. For $i \leqslant r - s$, $\chi^*(\{\alpha_i, \alpha_{i+1}, \ldots, \alpha_{i+s}\}) = 0$ so $\alpha_i' > \alpha_{i+1}'$, that is, $\alpha_1' > \cdots > \alpha_{r-s+1}'$. Let $i_1 < \cdots < i_{s-1} \leqslant r - s + 1$ be arbitrary. Then

$$\chi_{s-1}(\{\alpha_{i_1}', \ldots, \alpha_{i_{s-1}}'\}) = \chi_s(\{\alpha_{i_1}, \ldots, \alpha_{i_{s-1}}, \alpha_{i_{s-1}+1}\})$$

so that $\{\alpha_1', \ldots, \alpha_{r-s+1}'\}$ is monochromatic under χ_{s-1}. By induction

$$r - s + 1 \leqslant h_{s-1}(N(\alpha_1')) \leqslant h_{s-1}(N(\alpha_1))$$

[since h_{s-1} is monotone and $N(\alpha_1') \leqslant N(\alpha_1)$]. Setting

$$h_s(x) = \max[x + s - 2, e_{s-1}(x) + s - 1, h_{s-1}(x) + s - 1],$$

we have $r \leqslant h_s(N(\alpha_1))$ in all cases, completing the induction step. [In fact, $h_s(x) = e_{s-1}(x) + s - 1$ for $x \geqslant 3$ by a simple calculation. But our concern here is only to find some function h_s having the desired property.]

The apply the above colorings to sets of integers we shall define a correspondence between integers and ordinals. Let $\alpha < \varepsilon_0$, $n \in N$. We define an ordinal valued function $T = T^{\alpha, n}$, called the (n, α) translation function, inductively as follows. Set $T(n) = \alpha$. Assume that $T(m) = \beta$ has been defined. If $\beta > 0$ is not a limit ordinal set $T(m + 1) = \beta - 1$. If $\beta > 0$ is a limit ordinal set $T(m + 1) = \beta(m) - 1$. When $T(u) = 0$ terminate the definition. Define $U(n, \alpha)$ as the value u such that $T^{\alpha, n}(u) = 0$. Such a u exists, as otherwise $T(n), T(n + 1), \cdots$ would be an infinite descending sequence of ordinals.

Example. $\alpha = \omega^2$, $n = 5$. Then $T(5) = \omega^2$, $T(6) = 4\omega + 4$, $T(10) = 4\omega$, $T(11) = 3\omega + 9$, $T(20) = 3\omega$, $T(40) = 2\omega$, $T(80) = \omega$, $T(160) = 0$. $U(5, \omega^2) = 160$.

Property 8. $N[T^{\alpha, n}(m)] \leqslant \max[N(\alpha), m]$ for all α, n, m for which $T^{\alpha, n}(m)$ is defined. In particular, when $\alpha = \gamma_s$, $N[T^{\alpha, n}(m)] \leqslant m$.

The proof uses a simple induction on n, applying Property 2.

Property 9. For all $1 \leq \alpha < \varepsilon_0$, $n \in N$,

$$U(n, \omega^\alpha) = f_\alpha(n) ,$$

where f_α is the Ackermann function.

Proof. We use transfinite induction on α. For $\alpha = 1$

$$U(n, \omega) = 2n$$

trivially. Assume that Property 9 holds for all $\alpha' < \alpha$. If α is a limit ordinal then $\omega^\alpha(n) = \omega^{\alpha(n)}$, so $\omega^\alpha((n)) = \omega^{\alpha(n)}((n))$ and hence the (n, ω^α) and $(n, \omega^{\alpha(n)})$ translation functions are identical at $n + 1$. A single value of a translation function determines all succeeding values. Hence

$$U(n, \omega^\alpha) = U(n, \omega^{\alpha(n)}) = f_{\alpha(n)}(n) = f_\alpha(n) .$$

Now assume that $\alpha = \beta + 1$. We observe that, in general,

$$T^{(s+1)\omega^\beta, t}(i) = s\omega^\beta + T^{\omega^\beta, t}(i) , \qquad t \leq i \leq U(t, \omega^\beta) = f_\beta(t) ,$$

as the $s\omega^\beta$ term remains fixed in defining the translation function. By induction on s

$$U(t, s\omega^\beta) = f_\beta^{(s)}(t) .$$

The $(t, \omega^{\beta+1})$ and $(t, t\omega^\beta)$ translation functions are identical at $t + 1$. hence

$$U(t, \omega^{\beta+1}) = U(t, t\omega^\beta) = f_\beta^{(t)}(t) = f_{\beta+1}(t) ,$$

completing the induction.

Our preliminaries are complete. Now consider the following statement:

$P(s, m, n)$: $n > m$, and if $[m, n]^{s+1}$ is $(2s - 1)$-colored there exists a monochromatic set X such that $|X| \geq h_s(\min(X))$.

Property 10. The statement $P(s, m, n)$ is expressible in Peano arithmetic.

Since this is not a book on logic, we dismiss Property 10 as "obvious"—as indeed it is.

Property 11. $\forall_s \forall_m \exists_n P(s, m, n)$.

The proof is a simple application of the Compactness theorem, identical to the proof of (PH) at the beginning of this section.

Property 12. $P(s, m, n)$ is false for $n \leqslant f_{\gamma_{s-1}}(m)$.

Proof. Let T be the $(m, \gamma_s(m))$ translation function. $T(x)$ is defined for $m \leqslant x \leqslant U(m, \gamma_s(m)) = U(m, \gamma_s) = f_{\gamma_{s-1}}(m)$, hence for $m \leqslant x \leqslant n$. Recall that $x < y$ implies $T(x) > T(y)$. Also recall that $N(T(x)) \leqslant x$ (Property 8). Let χ_s be as previously defined. Define

$$\chi'_s : [m, n]^{s+1} \to \{0, 1, \ldots, 2s - 2\}$$

by

$$\chi'_s(\{x_1, \ldots, x_{s+1}\}) = \chi_s(\{T(x_1), \ldots, T(x_{s+1})\})$$

(i.e., identify $x \in [m, n]$ with $T(x) \in [\gamma_s]$). Let $X = \{x_1, \ldots, x_{s+1}\}_<$ be monochromatic under χ_s. Then $T(X) = \{T(x_1), \ldots, T(x_{s+1})\}_>$ (with order reversed) is monochromatic under χ_s. Hence

$$u = |X| = |T(X)| \leqslant h_s(N(T(x_1)))$$
$$\leqslant h_s(x_1) ,$$

since h_s is monochromatic; that is, χ'_s gives a counterexample to the statement $P(s, m, n)$.

Finally, for ease of expression, define

$$P(s, t) \stackrel{\text{def}}{=} P(s + 1, s, t) .$$

Then $P(s, t)$ is false, for $t \leqslant f_{\gamma_s}(s) = f_{\varepsilon_0}(s)$. P is provably recursive as the veracity of $P(s, t)$ can be determined by checking all $(2s + 1)$-colorings of $[s\ t]^{s+2}$. Combining our remarks, we find that P satisfies (PH0), (PH1), (PH2), and (PH3) so the statement

(PH′)
  For all s there exists t so that, if $[s, t]^{s+2}$ is $(2s + 1)$-colored, there exists a monochromatic set X such that $|X| \geqslant h_{s+1}(\min(X))$

is true for the integers but not provable in Peano arithmetic.

Notes. Paris and Harrington [1977] give a model-theoretic proof that (PH) is undecidable in Peano arithmetic. Our approach is based on Ketonen and Solovay [1981]. G. Kreisel [1952] gives the proof theory result. Spencer [1983] gives an expository overview of these results.

6.4 THE INFINITE

In this book we have purposely restricted our attention to finite Ramsey results, proving infinite results only to show, via a Compactness argument, a finite theorem. However, there is an enormous literature on Infinite Ramsey theorems per se. In this section we mention a few esthetically appealing, self-contained results from this literature. We assume the Axiom of Choice throughout.

Let α, β be cardinals. We define

$$\beta \rightarrow (\alpha)^2$$

if, whenever $|A| \geq \beta$ and $[A]^2$ is 2-colored, there exists $B \subset A$, $|B| \geq \alpha$, with $[B]^2$ monochromatic.

Let c denote the cardinality of the set of real numbers.

Theorem 1. $c \nrightarrow (c)^2$.

Proof. Let $<$ be the usual ordering of R and $<'$ a well ordering. We 2-color $[R]^2$ by

$$\chi(\{x, y\}_<) = \begin{cases} \text{red} & \text{if } x <' y, \\ \text{blue} & \text{if } y <' x. \end{cases}$$

Assume that $[S]^2$ is red. Then S is well ordered by $<$. For all $x \in S$ (except the maximal element m, if one exists) there exists $x^+ \in S$, $x < x^+$ so that $S \cap (x, x^+) = \varnothing$. Let $A_n = \{x: x^+ - x > n^{-1}\}$. Clearly, A_n is countable so $S = \bigcup_{n=1}^{\infty} A_n \cup \{m\}$ is also countable. Similarly, if $[S]^2$ is blue then S is countable. Thus we have proved the stronger result $c \nrightarrow (\omega_1)^2$, where ω_1 is the first uncountable cardinal.

Theorem 2. For all α there exists β so that $\beta \rightarrow (\alpha)^2$. In particular, if α is an infinite cardinal then

$$(2^\alpha)^+ \rightarrow (\alpha)^2.$$

Proof. Let $|A| = (2^\alpha)^+$, and fix a 2-coloring $\chi: [A]^2 \to \{0, 1\}$. As only the cardinality of A is of importance, set

$$A = \{0, 1, \cdots\} = \{\delta : \delta < (2^\alpha)^+\},$$

the set of ordinals up to $(2^\alpha)^+$. By transfinite induction, define for each $i \in A$ a well-ordered sequence $S(i)$ of 0's and 1's. Define $S(0)$ to be the null sequence. Assume that $S(i)$ has been defined for $i < n$. The first term of $S(i)$, that is, $(S(i))(1)$, is defined by

$$(S(i))(1) = \chi(0, i).$$

Now assume that $(S(i))(j)$ has been defined for $j < t$. If, for some $\alpha < i$, the sequence $S(a)$ is equal to the portion of $S(i)$ already constructed then we define

$$(S(i))(t) = \chi(a, i). \tag{$*$}$$

If no such a exists we terminate the sequence $S(i)$. The process is illustrated in Fig. 6.1. To each $i \in A$ is associated a distinct sequence $S(i)$, for if $i < j$ and $S(i) = S(j)$ then the sequence $S(j)$ was terminated when it should not have been. Note also that this implies that the a in $(*)$ is uniquely defined if it exists. There are at most 2^α well-ordered zero-one sequences of length $< \alpha$. Hence for some t the sequence $S(t)$ is of length α. For all $i < \alpha$ there exists a_i so that $S(a_i)$ forms the first i terms of $S(t)$. Let $i < j < \alpha$. Then

$$\chi(a_i, a_j) = (S(j))(i + 1) = (S(t))(i + 1),$$

independently of j (e.g., $\{0, 2, 3, 6, 7\}$ in Fig. 6.1). On $A' = \{a_i\}_{i < \alpha}$ we define a point coloring χ' by

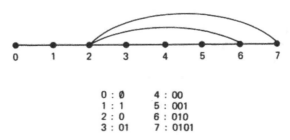

0 : ∅	4 : 00
1 : 1	5 : 001
2 : 0	6 : 010
3 : 01	7 : 0101

Figure 6.1 Vertex-Sequence correspondence.

$$\chi'(a_i) = \chi(a_i, a_j), \qquad i < j.$$

If a set of cardinality α is partitioned into two parts one part must have cardinality α. Let $B \subseteq A'$, $|B| = \alpha$, B monochromatic under χ'. Then B is monochromatic under χ. This completes the proof. In fact, $(2^\alpha)^+ \rightarrow (\alpha^+)^2$ may be proved by a suitable modification of the proof just given.

Erdös, Hajnal, and Rado [1965] consider Ramsey relations for cardinals in great detail.

There are many combinatorial questions involving Ramsey numbers for countable ordinals. We write

$$\gamma \rightarrow (\alpha, \beta)$$

(α, β, γ ordinals) if, whenever $[\gamma]^2$ is red-blue colored, there exists either a red $[S]^2$, S of order type α, or a blue $[T]^2$, T of order type β. We give one relatively simple result in this area (see also Section 6.3).

Theorem 3. $\omega^2 \rightarrow (\omega^2, m)$ for all $m < \omega$.

Proof. Let χ be a red-blue coloring of $[\omega^2]^2$. We define an 8-coloring χ' of $[N]^4$ by

$$\chi'(\{a, b, c, d\}_<) = (\chi(a\omega + b, c\omega + d), \chi(a\omega + c, b\omega + d),$$

$$\chi(a\omega + d, b\omega + c)).$$

There exists an infinite $A \subseteq N$, monochromatic under χ'. For convenience we relabel so that $A = N$.

If all $\chi(a\omega + b, cw + d) = $ blue then $\{\omega + 2, 3\omega + 4, \dots, (2m - 1) \omega + 2m\}$ is a blue K_m.

If all $\chi(a\omega + c, b\omega + d) = $ blue then $\{\omega + (m + 1), 2\omega + (m + 2), \dots, m\omega + 2m\}$ is a blue K_m.

If all $\chi(a\omega + d, b\omega + c) = $ blue then $\{\omega + 2m, 2\omega + (2m - 1), \dots, m\omega + (m + 1)\}$ is a blue K_m.

If none of the above, $\chi' = $ (red, red, red) so that $\chi(w\omega + x, y\omega + z) = $ red for all distinct w, x, y, z with $w < x$, $y < z$. For each prime p (this is only a convenience to ensure distinctness), set $S_p = \{pw + p^n : n \geq 2\}$. For each p either S_p contains a blue K_m (in which case we are done) or an infinite red $T_p \subseteq S_p$. If the latter holds for all p then $T = \cup T_p$ is a blue set of order type ω^2.

An example of a truly difficult problem in this area is the relation

$$\omega^\omega \rightarrow (\omega^\omega, 3)^2 \ .$$

This was proved by Chang [1972]; his proof was simplified by Larson [1973]. Erdös and Hajnal [1971] give numerous problems involving ordinals.

We now consider colorations of all finite subsets of a set A. We call $B \subseteq A$ *well-colored* if $[B]^i$ is monochromatic for all integers i. (We could not expect $[B]^{<\omega}$ to be monochromatic since one could, for example, color $[A]^2$ red and $[A]^3$ blue.) Ramsey, in his original paper (Chapter 1, Theorem 8), showed that, for all k, r, there exists n so that if $|A| \geq n$ and $[A]^{<\omega}$ is r-colored there exists a well-colored $B \subseteq A$, $|B| = k$. The natural generalization to infinite B is false. Define a 2-coloring of $[N]^{<\omega}$ by

$$\chi(X) = \begin{cases} 0 & \text{if } |S| \in S, \\ 1 & \text{if } |S| \notin S. \end{cases}$$

If B is infinite, let $b_1 \in B$. There are subsets $X, Y \subseteq B$, $|X| = |Y| = b_1$ with $b_1 \in X$ and $b_1 \notin Y$. Thus $\chi(X) \neq \chi(Y)$ so B is not well colored.

Let us call A *small* if there exists a 2-coloring of $[A]^{<\omega}$ so thate there is no infinite well-colored $B \subseteq A$,

Theorem 4. If A is small, 2^A is small.

Proof. Let χ be a 2-coloring of $[A]^{<\omega}$ with no well-colored infinite B. Well-order A by $<$. For distinct $X, Y \subset A$ set

$$\overline{XY} = \min X \triangle Y \ ,$$

where $\triangle$ denotes symmetric difference. Order 2^A lexicographically, setting $X < Y$ iff $\overline{XY} \in X$. If $X < Y < Z$ then $\overline{XZ} = \max(\overline{XY}, \overline{YZ})$ and $\overline{XY} \neq \overline{YZ}$. More generally, if $X_1 < \cdots < X_n$ then $\overline{X_1 X_n} = \max X_i X_{i+1}$.

Now let us define χ' on $[2^A]^{<\omega}$. Let $\{X_1, \ldots, X_n\}_< \subset 2^A$. Set $\varepsilon_i = \overline{X_i X_{i+1}}$, $1 \leq i < n$. If the ε_i are monotonically increasing or decreasing, set

$$\chi'(\{X_1, \ldots, X_n\}) = \chi(\{\varepsilon_1, \ldots, \varepsilon_{n+1}\}) \ .$$

Otherwise, let χ' be arbitrary.

Suppose that B were an infinite well-colored subset of 2^A. If $\{X, Y, Z\}_< \subset B$ and $\overline{ZY} < \overline{YZ}$ we call $\{X, Y, Z\}_<$ mono-up; if $\overline{XY} >$

$\overline{YZ}$, mono-down. By Ramsey's theorem there is an infinite $C \subseteq B$ where all $\{X, Y, Z\}_< \subset C$ have the same orientation, say mono-up (mono-down is similar).

An infinite ordered set may be shown to contain a subset order isomorphic to either N or $(-N)$. Restrict C to such a subset, say

$$C = \{X_1, X_2, \cdots\}_<$$

[order type $(-N)$ is similar]. Set $\varepsilon_i = \overline{X_i X_{i+1}} \in A$ so that $\varepsilon_1 < \varepsilon_2 < \cdots$. The ε's reflect the behavior of the X's, since

$$\chi(\{\varepsilon_{i_1}, \ldots, \varepsilon_{i_n}\}) = \chi(\{X_{i_1-1}, X_{i_1}, X_{i_2}, \ldots, X_{i_n}\})$$

for all $2 \le i_1 < \cdots < i_n$. Thus $\{\varepsilon_i : 2 \le i\}$ would be well-colored under χ, contradicting our assumption. Thus no infinite well-colored B exists, and 2^A is small.

Theorem 5. Let $A = \cup_{\alpha \in I} A_\alpha$, where the A_α are pairwise disjoint. Assume that I is small and all A_α are small. Then A is small.

Proof. Let $\chi_\alpha : [A_\alpha]^{<\omega} \to [2]$ denote the coloring showing A_α to be small, and $\chi^* : [I]^{<\omega} \to [2]$ denote the coloring showing I to be small. We define χ on $[A]^{<\omega}$ as follows. Let $X = \{x_1, \ldots, x_s\} \subseteq A$. If all x_i are in the same A_α set $\chi(X)$. If $x_i \in A_{\alpha_i}$ and the α_i are distinct set $\chi(X) = \chi^*(\{\alpha_1, \ldots, \alpha_s\})$. Otherwise let χ be arbitrary.

Let $B \subseteq A$ be infinite. There exists an infinite $C \subseteq B$ so that either all $x \in C$ are in the same A_α or all $x \in C$ are in distinct A's. On C, χ reflects either χ_α or χ^* so that C, and hence B, are not well-colored. Hence A is small.

Are there any sets A that are not small? If the answer is yes there will be a cardinal β_0 that, is the smallest cardinality of a set A that is not small. β_0 is called the first Erdös cardinal. Our previous theorems have shown that $\beta_0 \neq \omega$ and that:

(i) if $\alpha < \beta_0$ then $2^\alpha < \beta_0$;
(ii) if $\lim_{\gamma \in I} \alpha_\gamma = \alpha$, where $\alpha_\gamma < \beta_0$ for all $\gamma \in I$ and I is a well-ordered set with $|I| < \beta_0$, then $\alpha < \beta_0$.

In the jargon of the set theorists, β_0 is power set inaccessible and limit inaccessible; β_0 is what is called an inaccessible cardinal. Do inaccessible cardinals exist? One cannot prove their existence from the usual axioms of set theory, for, in a nutshell, if a smallest inaccessible cardinal α_0 existed the family of ordinals $\alpha < \alpha_0$ would provide a model for set theory in which no inaccessible cardinals exist. It appears that the existence of β_0 does not contradict the usual axioms of set theory—but only for the heuristic reason that no contradiction has been found. In fact, the existence $\cdots$ but enough. We have strayed into the arcane world of "large cardinal axioms," where questions may be answered by yes, no, and various shades of maybe. This is a finite book on finite mathematics. We choose to stop here.

References

Ajtai, M., Komlós , J., and Szemerédi, E. (1980), A Note on Ramsey Numbers, *J. Comb. Th. (A)* **29**, 354–360.

Baumgartner, J. E. (1974), A Short Proof of Hindman's Theorem, *J. Comb. Th. (A)* **17**, 384–386.

Beck, J. (1978), On 3-Chromatic Hypergraphs, *Discrete Math* **24**, 127–137.

Behrend, F. A. (1946), On Sets of Integers Which Contain No Three in Arithmetic Progression, *Proc. Nat. Acad. Sci.* **23**, 331–332.

Bergelson, V. and Hindman, N. (in press), Nonmetrizable Topological Dynamics and Ramsey Theory, *Trans. Amer. Math. Soc.*

Berlekamp, E. R. (1968), A Construction for Partitions Which Avoid Long Arithmetic Progressions, *Can. Math. Bull.* **11**, 409–414.

Burkill, H. and Mirsky, L. (1973), Monotonicity, *J. Math. Anal. Appl.* **41**, 391–410.

Burling, J. P. and Reyner, S. W. (1972), Some Lower Bounds on the Ramsey Numbers $n(k, k)$, *J. Comb. Th. (B)* **13**, 168–169.

Burr, S. A. (1974), Generalized Ramsey Theory for Graphs—A Survey, in *Graphs and Combinatorics*, Springer, pp. 52–75.

Burr, S. A. and Erdös, P. (1976), External Ramsey Theory for Graphs, *Utilitas Math.* **9**, 247–258.

Burr, S. A., Erdös, P., and Spencer, J. H. (1975), Ramsey Theorems for Multiple Copies of Graphs, *Trans. Amer. Math. Soc.* **209**, 87–99.

Burr, S. A., Faudree, R. J., and Schelp, R. H. (1977), On Ramsey—Minimal Graphs, Proceedings of the Eighth Southeastern Conference on Combinatorics, Graph Theory and Computing, *Utilitas, Congressus Numerantium XIX* 115–124.

Chang, C. C. (1972), A Partition theorem for the Complete Graph on ω^ω, *J. Comb. Th. (A)* **12**, 396–452.

Cates, M. L. and Hindman, N. (1975), Partition Theorems for Subspaces of Vector Spaces, *J. Comb. Th. (A)* **18**, 13–25.

Choi, S. L. G. (1971), On the Density of Certain Integer Sequences, *Proc. London Math. Soc.* **23**, 565–576.

Chung, F. R. K. and Graham, R. L. (1975), On Multicolor Ramsey Numbers for Complete Bipartite Graphs, *J. Comb. Th. (B)* **18** 164–169.

Chvátal, V. (1969), On Finite Polarized Partition Relations, *Can. Math. Bull.* **12**, 321–326.

——— (1970), Some Unknown van der Waerden Numbers, *Combinatorial Structures and Their Applications*, Gordon & Breach, pp. 31–33.

——— (1977), Tree-Complete Graph Ramsey Numbers, *J. Graph Th.* **1**, 93.

Chvátal, V. Harary, F. (1972), Generalized Ramsey Theory for Graphs, III. Small Off-Diagonal Numbers, *Pac. J. Math.* **41**, 335–345.

Comfort, W. (1977), Ultrafilters: Some Old and Some New Results, *Bull. Amer. Math. Soc.* **83**, 417–455.

DeBruijn, N. G. and Erdös, P. (1951), A Color Problem for Infinite Graphs and a Problem in the Theory of Relations, *Nederl. Akad. Wetensch. Proc. Ser. A.* **54**, 371–373.

Deuber, W. (1973), Partitioned und Lineare Gleichungssysteme, *Math. Zeit.* **133**, 109–123.

—— (1975a), Partition Theorems for Abelian Groups, *J. Comb. Th. (A)* **19**, 95–108.

—— (1975b), Partitionstheoreme für Graphen, *Math. Helvetici* **50**, 311–320.

Dilworth, R. P. (1950), A Decomposition Theorem for Partially Ordered Sets, *Ann. Math.* **51**, 161–166.

Erdös, P. (1947), Some Remarks on the Theory of Graphs, *Bull. Amer. Math. Soc.* **53**, 292–294.

—— (1950), Some Remarks on Set Theory, *Proc. Amer. Math. Soc.* **1**, 127–141.

—— (1959), Graph Theory and Probability, *Can. J. Math.* **11**, 34–38.

—— (1961), Graph Theory and Probability, II, *Can, J. Math.* **13**, 346–352.

—— (1963a), On a Combinatorial Problem, *Nord. Mat. Tidskr.* **11**, 5–10.

—— (1963b), External Problems in Graph Theory, in *Theory of Graphs and Its Applications* (edited by W. T. Tutte), Publ. House Czechoslovak Acad. Sci. Prague, pp. 26–36.

—— (1964a), On a Combinatorial Problem, II, *Acta Math. Acad. Sci. Hung.* **15**, 445–447.

—— (1964b), On Extremal Problems of Graphs and Generalized Graphs, *Israel J. Math.* **2**, 183–190.

—— (1969), On a Combinatorial Problem, III, *Can Math. Bull.* **12**, 413–416.

—— (1973), in *The Art of Counting* (edited by J. Spencer), M.I.T. Press.

Erdös, P., Faudree, R. J., Rosseau, C. C., and Schelp, R. H. (1978), The Size Ramsey Number, *Per. Math. Hung.* **9**, 145–161.

Erdös, P. and Graham, R. L. (1975), On Partition Theorems for Finite Graphs, in *Infinite and Finite Sets* (edited by A. Hajnal, R. Rado, and V. T. Sós), North Holland, pp. 515–527.

Erdös, P., Graham, R. L., Montgomery, P., Rothschild, B. L., Spencer, J., and Straus, E. G. (1973), Euclidean Ramsey Theorems, I, *J. Comb. Th.* **14**, 341–363.

—— (1975a), Euclidean Ramsey Theorems, II, in *Infinite and Finite Sets* (edited by A. Hajnal, R. Rado, and V. T. Sós), North Holland, pp. 529–558.

—— (1975b), Euclidean Ramsey Theorems, III, in *Infinite and Finite Sets* (edited by A. Hajnal, R. Rado, and V. T. Sós), North Holland, pp. 559–584.

Erdös, P. and Hajnal, A. (1971), Ordinary Partition Relations for Ordinal Numbers, *Per. Math. Hung.* **1**, 171–185.

Erdös, P., Hajnal, A., and Rado, R. (1965), Partition Relations for Cardinal Numbers, *Acta Math. Acad. Sci. Hung.* **16**, 93–196.

Erdös, P. and Lovasz, L. (1975), Problems and Results on 3-Chromatic Hypergraphs and Some Related Questions, in *Infinite and Finite Sets* (edited by A. Hajnal, R. Rado, and V. T. Sós), North Holland, pp. 609–628.

Erdös, P. and Moon, J. W. (1964), On Subgraphs of the Complete Bipartite Graphs, *Can. Math. Bull.* **7**, 35–39.

Erdös, P. and Moser, L. (1964), A Problem on Tournaments, *Can. Math. Bull.* **7**, 351–356.

Erdös, P. and Rado, R. (1950), A Combinatorial Theorem, *J. London Math. Soc.* **25**, 249–255.

—— (1952), Combinatorial Theorems on Classifications of Subsets of a Given Set, *Proc. London Math. Soc.* **2**, 417–439.

—— (1956), A Partition Calculus in Set Theory, *Bull. Amer. Math. Soc.* **62**, 427–489.

Erdös, P. and Sós, V. T. (1969), Some Remarks on Ramsey's and Turan's Theorems, in *Colloquia Mathematica Societatis Janos Bolyai 4. Combinatorial Theory and Its Applications, Balatonfured, Hungary*, North Holland, pp. 395–404.

Erdös, P. and Spencer, J. (1974), *Probabilistic Methods in Combinatorics*, Academic Press.

Erdös, P. and Szekeres, G. (1935), A Combinatorial Problem in Geometry, *Composito Math.* **2**, 464–470.

—— (1962), On Some Extremum Problems in Elementary Geometry, *Eotvos Sect. Math.* **3–4**, 53–62.

Erdös, P. and Turan, P. (1936), On Some Sequences of Integers, *J. London Math. Soc.* **11**, 261–264.

Exoo, G. (1989), A Lower Bound for $R(5,5)$, *J. Graph. Th.* **12**, 97–98.

Folkman, J. (1970), Graphs with Monochromatic Complete Subgraphs in Every Edge Coloring, *SIAM J. Appl. Math.* **18**, 19–24.

—— (1974), Notes on the Ramsey Number $N(3,3,3,3)$, *J. Comb. Th.* (A) **16** 371–379.

Frankl, P. (1977), A Constructive Lower Bound for Some Ramsey Numbers, *Ars Comb.* **3**, 297–302.

Frankl, P., Graham, R. L., and Rödl, V. (1987), Induced Restricted Ramsey Theorems for Spaces, *J. Comb. Th.* (A) **44**, 120–128.

Frankl, P. and Rödl, V. (1986), All Triangles Are Ramsey, *Trans. Amer. Math. Soc.* **297**, 777–779.

Furstenberg, H. (1977), Ergodic Behavior of Diagonal Measures and a Theorem of Szemerédi on Arithmetic Progressions, *J. Anal. Math.* **31**, 204–256.

Furstenberg, H. (1981), *Recurrence in Ergodic Theory and Combinatorial Number Theory*, Princeton University Press.

Furstenberg, H. and Weiss, B. (1978), Topological Dynamics and Combinatorial Number Theory *J. Anal. Math.* **34**, 61–85.

Giraud, G. (1969), Majoration du Nombre de Ramsey Ternaire-Bicolore en (4, 4), *C. R. Acad. Sci. Paris* (A) **269**, 620–622.

—— (1973), Analyse Combinatoire, *C. R. Acad. Sci. Paris* (A) **276**, 1173–1175.

Gleason, A. M. and Greenwood, R. E. (1955), Combinatorial Relations and Chromatic Graphs, *Can. J. Math.* **7**, 1–7.

Gottschalk, W. H. (1951), Choice Functions and Tychonoff's Theorem, *Proc. Amer. Math. Soc.* **2**, 172.

Gowers, W. T. (2001), A New Proof of Szemerédi's Theorem, *Geom. Funct. Anal.* **11**, 465–588.

Graham, R. L. (1968), On Edgewise 2-Colored Graphs with Monochromatic Triangles and Containing No Complete Hexagon, *J. Comb. Th.* **4**, 300.

Graham, R. L., Leeb, K., and Rothschild, B. L. (1972), Ramsey's Theorem for a Class of Categories, *Adv. Math.* **8**, 417–433.

Graham, R. L. and Rothschild, B. L. (1974), A Short Proof of van der Waerden's Theorem on Arithmetic Progressions, *Proc. Amer. Math. Soc.* **42**, 385–386.

Graham, R. L. and Rödl, V. (1987), Numbers in Ramsey Theory, in *Surveys in Combinatorics* (edited by C. Whitehead), LMS Lecture Note Series 123, Cambridge University Press, pp. 111–153.

Grimstead, C. and Roberts, S. (1982), On the Ramsey Numbers $R(3,8)$ and $R(3,9)$, *J. Comb. Th. (B)* **33**, 27–51.

Guy, R. K. (1968), A Problem of Zarankiewicz, in *Theory of Graphs* (edited by P. Erdös and G. Katona), Academic Press, pp. 119–150.

——— (1969), A Many-Faceted Problem of Zarankiewicz, *Lecture Notes Math.* **110**, 129–148.

Guy, R. K. and Znam, S. (1969), A Problem of Zarankiewicz, in *Recent Progress in Combinatorics*, Academic Press, pp. 237–243.

Hales, A. W. and Jewett, R. I. (1963), Regularity and Positional Games, *Trans. Amer. Math. Soc.* **106**, 222–229.

Hilbert, D. (1892), Uber die Irreducibilitat Ganzer Rationaler Functionen mit Ganzzahligen Coefficienten, *J. Reine Angew Math.* **110**, 104–129.

Hindman, N. (1974), Finite Sums from Sequences within Cells of a Partition of N, *J. Comb. Th. (A)* **17**, 1–11.

Irving, R. (1973), On a Bound of Graham and Spencer for a Graph-Coloring Constant, *J. Comb. Th. (B)* **15**, 200–203.

——— (1974), Generalized Ramsey Numbers for Small Graphs, *Disc. Math.* **9**, 251–264.

Isbell, J. R. (1969), $N(4,4;3) \geqslant 13$, *J. Comb. Th.* **6**, 210.

Kalbfleisch, J. G. (1967), Upper Bounds for Some Ramsey Numbers, *J. Comb. Th.* **2**, 35–42.

——— (1971), On Robillard's Bounds for Ramsey Numbers, *Can. Math. Bull.* **14**, 437–440.

Kalbfleisch, J. G. and Stanton, R. C. (1968), On the Maximal Triangle-Free Edge-Chromatic Graphs in Three Colors, *J. Comb. Th.* **5**, 9–20.

Ketonen, J. and Solovay, R., (1981), Rapidly Growing Ramsey Functions, *Ann. of Math.* **113**, 267–314.

Kovari, T., Sós, V. T., and Turan, P. (1954), On a Problem of Zarankiewicz, *Colloq. Math.* **3**, 50–57.

Kreisel, G. (1952), On the Interpretation of Nonfinitist Proofs II, *J. Symb. Logic* **17**, 43–58.

Larson, J. A. (1973), A Short Proof of a Partition Theorem for the Ordinal ω^ω, *Ann. Math. Logic* **6**, 129–145.

The Mathematics of Paul Erdös, vol. I, 2nd edition, S. Butler, R. L. Graham, and J. Nesetril, eds., Springer, New York, 2013.

Milliken, K. R. (1975), Ramsey's Theorem with Sums or Unions, *J. Comb. Th. (A)* **18**, 276–290.

Mirsky, L. (1975), The Combinatorics of Arbitrary Partitions, *Bull. Inst. Math.* **11**, 6–9.

Moser, L. (1953), On Non-averaging Sets of Integers, *Can. J. Math.* **5**, 245–252.

——— (1960), On a Theorem of van der Waerden, *Can. Math. Bull.* **3**, 23–25.

——— (1970), Problem 170, *Can. Math. Bull.* **13**, 268.

Motzkin, T. S. and Straus, E. G. (1965), Maxima for Graphs and a New Proof of a Theorem of Turan, *Can. J. Math.* **17**, 533–540.

Nešetřil, J. and Rödl, V. (1976), The Ramsey Property for Graphs with Forbidden Complete Subgraphs, *J. Comb. Th.* (*B*) **20**, 243–249.

—— (1978a), A Simple Proof of Galvin-Ramsey Property of Finite Graphs and a Dimension of a Graph, *Disc. Math.* **23**, 49–55.

—— (1978b), The Structure of Critical Ramsey Graphs, *Colloq. Internat. C.N.R.S.* **260**, 307–308.

Paris, J. and Harrington, L. (1977), A Mathematical Incompleteness in Peano Arithmetic, in *Handbook of Mathematical Logic* (edited by J. Barwise), North Holland, pp. 1133–1142.

Promel, H. (1986), Partition Properties of q-Hypergraphs, *J. Comb. Th.* (*B*) **41**, 356–385.

Rado, R. (1933a), Verallgemeinerung Eines Satzes von van der Waerden mit Anwendungen auf ein Problem der Zahlentheorie, *Sonderausg. Sitzungsber. Preuss. Akad. Wiss. Phys.-Math. Klasse* **17**, 1–10.

—— (1933b), Studien zur Kombinatorik, *Math. Zeit.* **36**, 242–280.

—— (1936), *Some Recent Results in Combinatorial Analysis*, Congres International des Mathematiciens, Oslo.

—— (1943), Note on Combinatorial Analysis, *Proc. London Math. Soc.* **48**, 122–160.

—— (1949), Axiomatic Treatment of Rank in Infinite Sets, *Can. J. Math.* **1**, 337–343.

—— (1969), Some Partition Theorems, in *Colloquia Mathematica Societatis Janos Bolyai 4. Combinatorial Theory and Its Applications, Balatonfured, Hungary*, North Holland.

Radziszowski, S. (1994–2011), *Electronic Journal of Combinatorics,* Dynamic Surveys DS1, revisions #1 through #13.

Ramsey, F. P. (1930), On a Problem of Formal Logic, *Proc. London Math. Soc.* **30**, 264–286.

Ray-Chaudhuri, D. K. and Wilson, R. M. (1973), The Existence of Resolvable Block Designs, in *A Survey of Combinatorial Theory* (edited by J. Srivastava), North Holland, pp. 361–376.

Roth, K. (1952), Sur Quelques Ensembles d'Entiers, *C. R. Acad. Sci. Paris* **234**, 388–390.

—— (1953), On Certain Sets of Integers, *J. London Math. Soc.* **28**, 104–109.

—— (1954), On Certain Sets of Integers, II, *J. London Math. Soc.* **29**, 20–26.

—— (1967), Irregularities of Sequences Relative to Arithmetic Progressions, II, *Math. Ann.* **174**, 41–52.

Salem, R. and Spencer, D. C. (1942), On Sets of Integers Which Contain No Three Terms in Arithmetic Progression, *Proc. Nat. Acad. Sci.*, **28**, 561–563.

Sanders, J. (1969), A Generalization of Schur's Theorem, Dissertation, Yale University.

Schur, I. (1916), Uber die Kongruenz $x^m + y^m \equiv z^m$ (mod p), *Jber. Deutsch. Math.-Verein.* **25**, 114–117.

Seidenberg, A. (1959), A Simple Proof of a Theorem of Erdös and Szekeres, *J. London Math. Soc.* **34**, 352.

Shelah, S. (1988), Primitive Recursive Bounds for van der Waerden Numbers, *J. Amer. Math. Soc.* **1**, 683–697.

Skolem, T. (1933), Ein Kombinatorische Satz mit Anwendung auf ein Logisches Entscheidungsproblem, *Fundamenta Math.* **20**, 254–261.

Spencer, J. (1975), Ramsey's Theorem—A New Lower Bound, *J. Comb. Th.* (*A*) **18**, 108–115.

—— (1977), Asymptotic Lower Bounds for Ramsey Functions, *Disc. Math.* **20**, 69–76.

—— (1979), Ramsey's Theorem for Spaces, *Trans. Amer. Math. Soc.* **249**, 363–371.

Spencer, J. H. (1983), Large Numbers and Unprovable Theorems, *Amer. Math. Monthly* **90**, 669–675.

Straus, E. G. (1975), A Combinatorial Theorem in Group Theory, *Math. Computation* **29**, 303–309.

Szemerédi, E. (1969), On Sets of Integers Containing No Four Elements in Arithmetic Progression, *Acta Math. Acad. Sci. Hung.* **20**, 89–104.

────── (1975), On Sets of Integers Containing No k Elements in Arithmetic Progression, *Acta Arith.* **27**, 199–245.

Taylor, A. D. (1976), A Canonical Partition Relation for Finite Subsets of ω, *J. Comb. Th.* (*A*) **21**, 137–146.

Turan, P. (1941), Egy Grafelmeleti Szelsoertek-Feladatrol, *Math. Phys. Lapok* **48**, 436–452.

────── (1954), On the Theory of Graphs, *Colloq. Math.* **3**, 19–30.

van der Waerden, B. L. (1927), Beweis einer Baudetschen Vermutung, *Nieuw Arch. Wisk.* **15**, 212–216.

────── (1971), How the Proof of Baudet's Conjecture Was Found, in *Studies in Pure Mathematics* (edited by L. Mirsky), Academic Press, pp. 251–260.

Veech, W. A. (1977), Topological Dynamics, *Bull. Amer. Math. Soc.* **83**, 775–830.

Walker, K. (1971), An Upper Bound for the Ramsey Number $M(5,4)$, *J. Comb. Th.* (*A*) **11**, 1–10.

Zarankiewicz, K. (1951), *Colloq. Math.* **2**, 301.

Index

WILEY SERIES IN
DISCRETE MATHEMATICS AND OPTIMIZATION